ELECTRIC REVOLUTION

Myths & Truths about Electric Vehicles and Climate Disaster

Aidan McClean

First Published 2022

Website: aidanmcclean.com

This edition published in 2022

ISBN 978-99987-902-1-6

9 8 7 6 5 4 3 2 1

Type Setting: Stephen Maguire
Cover Artwork: Laura Crotti

09.22

For my children

Éabha, Conor, Sean, Lucy, and Max

...

and for the support & encouragement thank you Gabriela, Mam and Dad

Contents

FOREWORD

Ironically, I was introduced to Aidan McClean by a fellow petrolhead and supercar owner and enthusiast, who had invested in Aidan's revolutionary all-electric, all-digital car rental company, UFODRIVE.

Despite my friend's passion for fast, noisy, internal combustion engine-powered sports cars, he could obviously see the future for electric vehicles and Aidan's vision for making them accessible to the car rental market and beyond.

I've been similarly obsessed with performance cars ever since I was a toddler and have been fortunate enough to have owned, driven and written about a wide variety of them over the last 30-odd years. Although I've driven a few electric vehicles over my automotive journalism career, I've never considered them the keystone element in the ever more urgent fight against climate change.

However, this book has opened my eyes to a) the wilful, knowing the destruction that the fossil fuel industry has wreaked on our environment, climate and health over the last 100 years and b) the wider effect that switching to electric vehicles will have in shifting us to a cleaner, more sustainable tomorrow.

There is still so much misinformation about electric vehicles, both deliberately disseminated and born out of simple ignorance, that it is refreshing to read a well-researched and reasoned book delving into the reality behind the myths.

ELECTRIC REVOLUTION isn't preachy or dogmatic, it's intelligent, factual and impassioned. Has it converted me to immediately refuse the keys to the petrol-powered works of automotive art I've loved my whole life?

Not quite yet, but I am now absolutely willing to accept that my next "daily driver" may be electric and in fact, I'm ready to embrace our electrically powered future.

Viva La Revolution Electrique!

David Yu

Chief Operating Officer YUnique Marketing Ltd, Automotive Journalist, Contributing Writer Evo Magazine 1999 onwards.

INTRODUCTION

If you just bought a gasoline car it's likely the last time you ever will.

Electric cars came before the internal combustion engine. At the beginning of the last century, we were on a path to clean electric mobility, but then we took a massive wrong turn. The discovery of oil, and the gigantic hydrocarbon industry that followed, sealed the fate of our early electric cars, and sent the internal combustion engine on an incredible journey. This journey gave humanity mobility, freedom, and great prosperity, but also a poisoned planet. The oil industry and the internal combustion focused automakers have played a leading role in the wilful destruction of our world, our air, our climate and possibly even threatening our survival on this rock.

Less than 20 years ago, the electric car sparked back into life with the obscure and little-known launch of Tesla Motors. Since then, the automotive industry has been turned on its head. Every indicator now confirms that we are in the middle of an electric revolution that reaches beyond just cars. Most people are unaware of the rapidly accelerating pace of this revolution and what it will mean for them. The revolution is about to go exponential.

Whilst it is still mostly misunderstood, the electric car and the parallel green revolution is going to bring fundamental change and a huge opportunity. The electric car revolution does not solve all our climate problems, but it is a big catalyst, not just for cars, but also for the entire energy industry upon which we developed society for the past 200 years. The way we moved, the way we produce electricity and the way we consumed in the 20th century will be changed beyond recognition by the mid-21st century; almost no area of life will remain untouched.

Time is the one ingredient that has made life possible on earth; our world changed slowly before we came along. We marvel at the universe and our solar system, but we, human beings, and what we've achieved, are the great wonder. However, how we got here cannot be how we go forward. Change, radical change, is an imperative and now.

This change is going to be an incredibly positive event in human history, we just need to make it happen. To have a chance at some form of stable future, humanity must cut greenhouse gas emissions by almost 8% per year for the next decade. We still have time for a second chance and the electric car revolution is part of this solution. Often you are not aware you are in the middle of a revolution until you look back and realize it's happened. We are in the middle of a revolution that is going to have a profound impact on almost every aspect of your life, but the good news is, it's mostly going to be a positive impact. I began a journey over five years ago. This journey has brought me from a near-total oblivious lack of awareness about the air we breathe, the state of our climate and the path we're on, to embracing the bright alternatives that are available right now. My journey started with an electric car test drive that blew my mind about the possibilities, I am not even a car guy and drove big, heavy, polluting SUV diesels for years. What I didn't realize is that my journey started decades before that fateful test drive.

I'm an outdoors guy, my lifelong sport was competitive cycling for much of the first half of my adult life. I was mostly unaware of dangerous air quality, although I was always consciously aware of the fumes I would regularly inhale when out for a bike ride, especially whenever a car drove past me. As you will read later, the penny finally dropped the day after I first drove an electric car and was later out on my bike, only to inhale a lung full of diesel fumes. I quit my comfortable corporate life and embarked on a new direction.

Finding a way to accelerate us towards clean electric mobility became an obsession. Together with my co-founder, Renaud Marquet, we created the world's first all-electric, completely digital car rental company. We wanted the customer experience to be completely out of this world and so my partner, Gabriela, suggested we named it UFODRIVE. This book is not about the story of UFODRIVE, maybe -if we're lucky- we will get to tell that story someday. This story is about how we humans got ourselves into this mess, who's to blame, why it is going to change rapidly and explaining how we have a great green, clean future ahead of us.

In the first years of my electric journey, I was astonished that everyone I met, from all walks of life, had already formed an opinion about electric cars. Most of these views were based on a lack of fact, misinformation and in some cases damned, deliberate lies. I didn't write this book to blindly try and convince you of the merits of electric cars, or from an environmentalist's standpoint, which I don't regard myself as having. I wrote this book to give everyone an understanding of how we got here, what's going on and what it's going to mean. In writing and researching this book I have been astonished at just how little I knew about the quality of our air, the dangers to our planet and the amazing work taking place to try to rectify our past ways through technology. I am going to tell the story of our world, the air we breathe and what's wrong with it, big oil's part in deceiving us and how the end of the fossil fuel age is closer than you think. I'm going to tell the story of how the rise of Tesla has led us past a tipping point in the automotive world and how the electric revolution is now unstoppable.

In service of the hundreds of times I have witnessed surprise on people's faces when I explain just what's happening and how electric cars work, I am also going to try deciphering fact from fiction about electric mobility; everything from charging to the incredible advances in battery technologies. I will cover how all this technology has spilt over beyond the electric car to accelerate a great green transition, bringing a bright, clean, safe electric future for all of us. I had no idea, until very recently, just what I was breathing into my body every day. I had no idea of the extent of our utter climate destruction, which has accelerated over the past 20 years, my part in this destruction and what the alternative could be.

Do you know the poison you breathe in every day? Do you know where it comes from and why you accept it? Do you know how these poisons got in the air, the role of big oil and the internal combustion engine? Air pollution causes seven million premature deaths per year. The Covid pandemic death rate has been made worse in many of the poorest countries by air pollution. Just shifting to electric vehicles could save thousands of lives per year. The downtime during lockdowns, for example in China, has been estimated to have probably saved more lives than the virus took.

With the rise of Tesla, the end of the fossil fuel and the internal combustion engine (ICE) age is accelerating rapidly. It's not just about electric cars, but an entire future generation driven by electric transport from scooters to flying taxis and even electric commercial aircraft. The electric revolution is happening quicker than you think. We are witnessing the birth of Generation Electric.

Tesla brought the electric car back to life and revolutionised the industry, making a total zero-emissions mobility future a real possibility. Electric cars still represent just a tiny 1% proportion of the 1.4 billion cars on the world's roads today, having a minuscule impact on CO_2 emissions — so far. However, every major automaker has now announced its intention to make the electric shift, some faster than others. The pace of change, not just for automakers, but with battery technology, clean energy, carbon storage and connected technologies means that clean air mobility worldwide is now possible. These are not just nice new cool toys; they are becoming an absolute necessity.

In 2021, a major new study by the World Meteorological Organization

(WMO) predicts that we're going to already hit the danger threshold, set by the Paris Agreement, of the world, being more than 1.5 degrees hotter than pre-industrial levels by 2025. The implications of a runaway warming planet will be devastating for humanity. As you will see in this book, the earth has a natural ability to heal itself but there is a point at which it is too late and runaway warming will become unstoppable. We have dug, mined, drilled, and sucked fossil fuels out of the ground for almost 200 years and burned them, releasing vast quantities of greenhouse gases that were meant to be stored in the earth forever.

Despite all our incredible technological advancements over the last 150 years, in many ways how we get energy and move about has barely evolved. Until now, we have been in the technological dark ages when it comes to utilizing the natural energy sources available to us. Instead of accelerating the advancements of clean electric cars and energy technologies, we have been mostly subsidising the fossil fuel industry at the expense of the climate - This must stop. The world's top oil companies made around $1 trillion in profit in the past decade and devoted just 3% of their $100 billion capital investment to low-carbon technologies.

My eyes have been opened to the boundless opportunities the electric and green revolution offers us all. I have had the privilege of meeting hundreds of people over these past five years, from the leaders of some of the world's largest automakers to scientists working on new clean energy technologies. I have also been privileged to witness thousands of individuals make the journey from internal combustion to electric, and with an overwhelmingly positive response. There is simply no going back and why would we want to? The future generation is an electric one, in a clean energy abundant world. Free from the spectre of a looming climate disaster, free to advance human society to accomplish great things, even beyond our world. I have broken down the story of electric mobility and clean energy into over 80 individual sections you can refer to in the contents. After you've read this book and the next time you walk down the street, and a nice car drives past, just look at the tailpipe and think about what you're breathing and how lucky we are to have this second chance at an electric future and how we must not waste it.

From the days of gasoline-powered cars and fossil fuel produced energy, we're moving to a clean bright and exciting future, but it won't be easy. The path to a carbon-neutral world, clean safe air and affordable abundant energy is fraught with hurdles and danger but we can get there. This great green transition also offers abundant opportunities to start again and reimagine our world. In this book, I'm going to try to explain just how we got here, where we're going and what it's going to mean.

We are now on an unstoppable journey to electrification and decarbonization and I am privileged, and grateful, to be able to share the story of our great transition and why the next generation is Generation Electric.

CHAPTER 1

BREATHE

The Earth is a living, breathing, adapting miracle spinning at roughly 1,000 miles per hour in the vacuum of space. It can self-heal, can repair the scars of earthquakes, volcanoes, and asteroid impacts. It's beautiful and has a singularly astounding trait – the ability to support life.

It's our home, and as far as we know, possibly harbouring the only active intelligent life in existence anywhere. The 'Pale Blue Dot' is a photograph of Earth taken on Feb. 14, 1990, by NASA's Voyager 1 space probe at 3.7 billion miles. Astronomer Carl Sagan famously described this iconic photograph in his 1994 book of the same name with the opening lines "Look again at that dot. That's here. That's home. That's us. On it everyone you love, everyone you know, everyone you ever heard of, every human being who ever was, lived out their lives."

THE OVERVIEW EFFECT

Out of over seven billion of us (and the 100 billion humans who have ever lived), less than 600 people have been in space. When the Apollo

8 mission, in 1968, first travelled far enough for the astronauts to be able to see the full sphere of the earth they experienced something dramatic and completely unprecedented. They called it The Overview Effect.

Suddenly seeing our entire world; beautiful, blue, thriving and alone in the vast emptiness of space awoke a new, unfamiliar emotion. Space is hostile, alien, and unable to support life. A human would die in less than 20 seconds if exposed to outer space. With an absence of air pressure, our blood cannot remain a liquid in space and would boil from within, a very unpleasant, horrific, and swift death. Earth's fine blue line, as first fully observed by the Apollo astronauts, is all we have. That thin blue line wrapped around the Earth is our atmosphere. Without it, life does not exist. It is delicate and needs to be protected and maintained, not destroyed. We are extremely vulnerable when viewed from a few hundred thousand kilometres in space. Our solitary, lonely world is unique.

There is no planet B. Our world is composed of the elements of the universe. Carbon is one of the most abundant of them. Nature has produced and stored away carbon over millennia until humans came along. We have released millions of years of stored carbon in a tiny fraction of Earth's lifetime. Humans are consuming the Earth and jeopardising our future survival. The Earth's population has more than doubled since man first walked on the moon in 1969. The impact of these extra four billion humans on our climate and air has been devastating.

Since the 1970s there has been a growing awareness of climate impact caused by human activity. With the establishment of organisations like Greenpeace and images of destroyed wildernesses, oil spills and air pollution, we have become aware of the damage that has been done, yet we continue apace. The Overview Effect has not reached many. Only now, as climate change is becoming more self-evident by the season, are we starting to realise and act. It may be too late. There is the possibility of survival and even thriving into the future, but only through immediate, significant action. The options are available, and life can be better, but first I will explain how we have poisoned the air we breathe and how the electric car can be one big part of the overall solution.

WHAT ARE YOU MADE OF?

It's mid-September, 10 am. I'm cycling on a smooth road, a tough 8% grade climb in the Ardennes Forest right on the Luxembourg-Belgium border. It's super quiet, cool with the early Autumn air and the sun peppering through the pine forest canopy. I can hear my deep breaths echo, with the distant faint sound of birds in the forest. It smells sweet, fresh with a hint of woody damp. I tilt down my tinted cycling glasses to look at this beauty with my own eyes - and it is a beauty - a forest in all its splendour - carpeted with bright green moss. It just feels so good to have this moment to myself and enjoy a small pleasure in our world. Then I hear it. The low drone of an oncoming car. It gets closer and louder. I'm all over the road, zig-zagging, trying to get up this tough climb so I pull close to right to let them drive past. It was an SUV. I didn't know the model exactly as it was moving fast, but it had four shiny exhausts. The driver very politely waited till the road was straighter, then accelerated around me. I couldn't see any smoke or steam or exhaust of any kind when he drove past, but I did instantly get the smell. When you are pushing hard up an 8-10% climb for 10 minutes or so your lungs have opened. Your heart rate is in the 90+ percentile and you simply can't just hold your breath! It's toxic. The smell and taste. It's difficult to describe the smell of diesel fumes exactly, I think it's somewhere between burning plastic, motor oil and cigarettes. The taste is metallic, iron mixed with burned toast. I'm not an environmentalist and never campaigned, or to be truthful, thought much about the environment, my air, or the planet. I often used to repeat a joke I heard somewhere from a comedian – "what's the environment ever done for me!" But that was then, that was my ignorance about what was going on around me. What my contribution to the planet is and how it affects me.

So, the smell and taste of a diesel car in the beautiful forest while out for a spin on my bike bothered me. It ruined that lovely moment – boohoo. A bit of a first-world problem, isn't it? The smell lingered a bit in the windless woods. I could smell and taste that diesel extract long after the car had vanished down the road. With no wind, it just lingered and slowly dissipated. Finally, it was gone. No more smell or trace of it - great I can breathe again, but where did it go? This question and a desire to go deeper into the origin and content of that smell drove me to investigate further and deeper in the months that followed. The smell didn't magically leave the planet, just because I couldn't notice it after a while. It stays with us. Forever. It just thins out to a point where our olfactory sensors can't detect it anymore.

It's partly an aerosol. This means the particle diameter of the exhaust emissions are comprised of a small enough mass that they can become suspended in the air for hours. Just like Covid-19. Like the perfectly still dust motes, you see suspended in mid-air when in a dark room with a slight gap in the curtains sending a sunbeam through. Some of the particles fall to the floor as they are bigger. This is known as Particulate Matter or PM. These PM particles have a certain mass and size which interacts with our planet, environment, and us in different ways. The infamous $PM_{2.5}$ or smaller is the diameter of diesel emissions that can cause cancers, respiratory infections, and death, especially for the vulnerable, like Covid-19 victims.

You put food and drinks in your mouth every day. You know what you like, what tastes good or doesn't. You also have an instinctive survival trait to avoid contaminated foods. Foul-smelling food or unattractive dark colouring will normally subconsciously make you avoid it. It's a basic survival instinct. You have a conscious choice - put it in your mouth or not. You don't have the same choice when it comes to the air you breathe - so suck it up...but at least know what you are putting into your body, via your lungs, every day.

Our atmosphere, the breathable part, is comprised of oxygen, 21%, nitrogen 78%. The remainder is primarily carbon dioxide and methane. These gasses, representing less than a tenth of a per cent may seem insignificant, yet they have a disproportionate effect on our existence. The mix, which has changed little over recent millennia enables life in abundance. The change of oxygen content in the atmosphere over eons, combined with the seasons has created the incredible diversity of animals and plants that comprise our ecosystem. Continuous change in adapting to Earth's atmospheric and temperature conditions created survival competition. Over millions of years, these changes allowed only the most adaptable animals to survive and thrive. 99.9% of all life that ever lived on Earth didn't make it - they are extinct. The predators of the world had to become smarter and smarter to survive and stay fed. This slowly made intelligence one of the most important tools of any species. Eventually, there were apes using tools, making clothes to keep warm and building shelters. Amongst all the mighty predators the planet had seen over the millennia, one of the least likely animals rose and came to dominate the planet - us. Now, in the present day, air pollution causes seven million premature deaths per year, more than double Covid-19's toll. The World Health Organisation (WHO) estimates direct exposure to diesel in its liquid form can cause serious kidney damage. According to the WHO, Diesel is a Group 1 carcinogen. Nitrogen Oxides (NOx) and PMs from diesel fumes, are responsible for 36,000 premature deaths a year in the UK alone, surpassing deaths from tobacco or obesity.

According to the American Cancer Society, experiments with diesel fume exposure in animals has been found to cause alteration to the DNA of cells and has a direct link to lung cancer. There have also been multiple links found between exposure to diesel fumes and increased cancers of the throat, larynx, and stomach. There are ongoing investigations to links between exhaust fumes and increased chances of leukaemia in children. There is no question that diesel and petrol fume exposure is a detriment to human and animal life. Dr Jonathan Grigg, from Doctors Against Diesel states "If you're going to design something that would effectively deliver a toxic substance into the lungs, you couldn't do better than the diesel soot particle." Eventually, all the PM from diesel fumes exhaust lands on the ground, typically within 2-7 meters, however, some of the PM can still travel hundreds of meters from the original emission. The forests' moss carpet is coated in it. Those dusty cars left in car parks for too long are mostly covered with diesel and petrol PM. Don't touch that dust! We don't just breathe exhaust PM we also eat it. PM lands on our crops and is washed into the soil to be consumed by plants, animals and by the top of the food chain, us. Like DDT, the infamous crop chemical so lambasted in the 1970s by the newly formed US Environmental Protection Agency, it accumulates in the environment and us. It's extremely toxic to the human body. It's cancer-causing. It's an unnatural thing.

What began as my curiosity into the content of this poisonous air that I breathe in every day has led me to discover the wider impact internal combustion cars, the oil industry and vested interests have had in bringing about our impending climate disaster. In researching air pollution and all its components and effects, I also saw the link the past 200 years of the oil age directly has to our overheating planet. In the chapters ahead I will explain the human damage air pollution causes and how our insatiable appetite for oil, fossil fuels and energy has driven our planet to the brink of climate disaster. From big oil, particulate matter to carbon dioxide and methane concentrations in the atmosphere, I will explain just how perilous the situation is.

WHAT HAVE WE DONE?

In December 2020, António Guterres, Secretary-General of the United Nations, declared that "the planet is broken" in a speech at Columbia University. The following is why.

Toxic gases, such as those from cars, would never naturally have been put back into the atmosphere. The planet is a living, self-healing rock covered with a thin layer of life clinging to the surface. Just the top half a kilometre of the earth's surface supports complex life, whilst at sea, it ranges from the surface to around 10 kilometres deep. The rest of our world is dead rock, magma and iron. Think of an apple's skin, a layer of life, even thinner, clings to the rock that is our world. Our atmosphere is 100km thick with the Kármán line being the imaginary barrier between us and space. Most of this 100km is not liveable. Above 18 km, the water contained in the human body and blood begins to boil as the atmospheric pressure is too low to keep blood from turning to gas. This is known as the Armstrong line. You would be long dead before you even reach this line, as once you go above 6 kilometres, or 20,000 feet, you begin to run out of air. Our atmosphere from space looks like a thin glowing wrapper around our globe. It's fine, delicate, and precious. Without it, we, and everything else alive doesn't exist.

Our atmosphere has layers from the Troposphere, where we live, through to the Exosphere. The gas composition of these layers changes to varying degrees. At sea level, it's mostly comprised of Nitrogen and Oxygen. As we go higher, Nitrogen and Oxygen thin dramatically. At the higher levels, such as the Stratosphere, Ozone becomes more prevalent. Ozone protects us from the sun's rays. In the 1980s, the destruction of Ozone caused by human-made Chlorofluorocarbons (CFC) gasses was a huge issue and humanity addressed it relatively successfully. The infamous 'hole in the sky' was prevented from growing too large at the poles before it impacted billions of people. This was due mainly to a concerted global campaign by governments which resulted in the 1985 Vienna Convention on the Protection of the Ozone Layer and the 1987, Montreal Protocol. The Ozone threat was backed up by irrefutable science with few contradicting conspiracy theories, possibly due to the lack of internet fuelled misinformation. The solution was, in part, to ban and eliminate CFC propellant from aerosols, deodorants, paint and other common household products. Still to this day, there is a natural and partially man-made ozone hole seasonally appearing at the poles. The consequence is, of course, a higher propensity for sunburns and human skin cancers. Coming together and solving that crisis is an example of reversing 50 years of human-made atmospheric damage. However, we have been burning oil and oil-based products for over 150 years and the impact has been far more devastating than the Ozone crisis but still solvable.

If we could instantly stop air pollution, would the Earth repair itself? It was incredible to see the waters of Venice clear up and fish return after just a few short months of human enforced self-lockdown during the Covid-19 Pandemic. This and the Ozone recovery, in just 30 years, gives hope that if we correct our ways quickly, the planet will right itself naturally. At least, this is what I used to think. The problem with air pollution and the burning of fossil-based products is the sheer scale and more importantly, the speed of climate destruction; it's just too fast for the planet to self-adapt and heal. Since the beginning of the industrial revolution, around 1750, we have burned and emitted 1.5 trillion tons of CO_2 into the atmosphere. Today we pump out more than three times the level of CO_2 per year compared to 1970. The United States has emitted more than any other country to date (400 billion tonnes since 1750). It's responsible for 25% of historical emissions alone. The production of crude oil itself also produces 1.7 gigatons (1.7 billion tonnes) of carbon dioxide or 5% of all emissions per year. This is produced from 9,000 oil fields in 90 countries. I'll cover this in the next chapter on Big Oil.

Life can adapt, given time. 99.9% of all life has gone extinct since the beginning of life some 3.5 billion years ago. The asteroid that killed off the dinosaurs polluted the air and darkened the skies so much that 76% of all species and nearly all the planet's vegetation was wiped out. The eruption of Krakatoa in the 13th and 15th centuries triggered the Little Ice Age. This was a century-long cold period known as a volcanic winter. The 'Thames Frost Fairs' took place during this period when Londoners could ice skate on the river. This was a single event from which the earth recovered once the dust eventually settled. If the Earth can recover from catastrophic atmosphere pollution caused by volcanoes and asteroids, it can recover from the massive scale air contamination we have produced, right? The answer is yes, it can, in time likely reinvent itself, but it will be too late for us. We will be the dinosaurs.

As with the dinosaurs, a marked change in CO_2 has always been incompatible with life on earth. All previous five extinction events involved massive increases in atmospheric carbon. Volcanoes take millions of years to produce the amount of atmospheric carbon we have produced in less than 150. We have effectively been burning millions of years' worth of living organisms all at once. The Earth has taken it on the chin. The heat trapped by our increasing CO_2 concentration has been largely absorbed by the planet. The ocean has been absorbing much of the excess heat masking the impact however it is reaching its limit. The ocean's inability to continue to absorb more heat has caused

Earth to lose its balance. The speed of temperature change is faster than in the last 10,000 years. Too fast for natural recovery. Fossil fuels are the cause. Fossil fuels are just that - fossils. The fossils of long-extinct plants and animals accumulated in the soil and rock of the earth over billions of years. Humans tend to measure the passage of time relative to our lifetimes. One or two generations behind is familiar to us. One hundred years ago is a long time, one thousand, ancient. That's not long in Earth time. We have been around for a blink of time in the history of life on Earth. The Earth adapts and self-heals over millennia, not years. It took over 4 million years for the Earth to re-stabilise after the dinosaur extinction event, known as the K-T event. What we've been doing for the past one hundred years is on par with the extinction event of the dinosaurs.

Volcanoes are often cited by climate deniers claiming we emit just a fraction of the CO_2 these natural wonders do. However, according to the US Geologic Survey, humans emit 60 times more CO_2 than all the currently active volcanoes per year. Fossil made oil was never meant to be extracted from deep within the Earth's crust and burned. Never meant to be put back into the atmosphere. The Earth quietly put all the hydrocarbon remnants from dead life away for safe storage, forever. We have dug up millions of years of accumulated carbon dioxide, methane and other gases and released and burnt them. We have woken the past and consciously pumped it into our atmosphere. By unbalancing millions of years of adaptive, automatic planet-wide air conditioning in just a fraction of Earth time, we have poisoned our world. Recovering from this scale of climate abuse is a gargantuan challenge. The speed of change is simply too fast. Just a few degrees of temperature change, combined with the higher concentrations of CO_2 in the atmosphere, has pushed the environment close to a tipping point of no return. The runaway chemical reaction of these gasses can become unstoppable to a point where any action we take will be too late. Just stopping all fossil fuel burning today could be too little too late. The damage has been done.

We don't perceive geological time. The temperature rises in the last century have been faster than at any other time in the planet's history.

According to the Intergovernmental Panel on Climate Change, the "scientific evidence for warming of the climate is unequivocal". The Earth does periodically warm and cool over millennia. There have been seven cycles of warming and cooling over the past 650,000 years. This is due to minor variations in Earth's orbit around the sun.

The current warming, however, is nothing natural. Most of the warming has occurred in the past 50 years. 2020 was the joint warmest year on record, along with 2016. From satellite ice sheet monitoring, Antarctic ice core drilling and rock analysis, scientists have proven global warming. The oceans absorb 90% of the increased heat. They are almost one degree warmer on average since man walked on the moon. It takes enormous energy to cause this level of increase. We are frogs in the slowly boiling water with our hand on the gas.

WHAT IS AIR POLLUTION?

If I told you the water you drink is anything but pure, you wouldn't drink it. If water bottles come with poison particle warning labels ranging anything above zero you wouldn't drink it - yet every day, everywhere, you accept toxic air to varying degrees. Your air is unclean. It's contaminated by man-made air pollutants. Airborne PM particles are small enough to enable them to become suspended in the air. They range in diameter, for example, PM10 is a particle measuring 10 micrometres, a human hair is between 17 - 180 micrometres thick. PM10 is relatively big for a particle and does not penetrate the fine membranes of the lungs or skin. A PM of 2.5 however is small enough to pass through our lung and soft tissue membranes, such as the lining of our mouths. PM of 10 or less is classed as inhale-able particles according to the US Environmental Protection Agency, EPA. $PM_{2.5}$ is produced from modern combustion engines.

The mass of PM is measured as the concentration of particles per cubic meter of air, expressed as (ug/m3). The World Health Organisation recommends a healthy limit of 24ug/m3. The EU limit is 50ug/m3 whilst it is 65ug/m3 in the United States. Air pollution is responsible for an estimated 7 million premature deaths per year according to the

WHO. In Europe, with stringent standards compared to some other geographies, that number is over 400,000 deaths per year. $PM_{2.5}$ is the major health hazard culprit, from human-generated sources such as transportation, manufacturing, and energy production. The 2020 Covid Pandemic has been estimated to have killed around a 1.2million in comparison, just one-sixth of the air pollution deaths.

The level of air pollution can also be measured using the international Air Quality Index (AQI). It may surprise you to learn just how polluted your city is on any given day. I recently became obsessed with measuring the AQI wherever I was in the world. You can easily do this with Apps such as 'Air Visual'. These apps collate various information from measuring stations all over the world. Wind and temperature conditions affect each day. You can download these apps and find out, for your location, the PM10 and $PM_{2.5}$ as well as Ozone (O3), Sulphur Dioxide (SO_2), Carbon Monoxide (CO) and Nitrogen Dioxide (NO_2) - expressed as parts per billion. The AQI calculates your air on a simple scale from good to hazardous. Wanting to go deeper and becoming more obsessed with my air, I purchased an Air Quality monitor. A small handheld device from a company called 'BREATHE'. There are lots of alternatives and no incentive was given to me to go for this one. I began experimenting.

This device, a relatively cheap 100 bucks or so, accurately measures $PM_{2.5}$ and states the AQI. My BREATHE monitor can produce super high readings when I burn an incense stick or activate a humidifier diffuser. Whilst those particles produce a scary rating for $PM_{2.5}$ concentrations, on the hazardous scale, it doesn't necessarily mean they are hazardous to your health. Concentrations of $PM_{2.5}$ for Diesel, for example, at about 80ug/m3 is defined as carcinogenic by the WHO. There have been near catastrophic air pollution concentration events resulting in 1,000s of hospitalisations in the past, such as Beijing in 2013 or Paris in 2016. I was in Los Angeles in February 2020, just before the Covid Pandemic exploded, and witnessed for myself the brown, yellow haze float over the city. What I discovered on my travels was very surprising. When cooking indoors with no open windows, my $PM_{2.5}$ shot up to 250, classified as dangerous. On a recent flight, I was pleasantly surprised to see the reading at 1! Thanks to multiple air filters on modern commercial aircraft, yes airplane air is clean! Next to a major road, the reading can reach 50+, or unhealthy for sensitive groups such as an asthma sufferer. Air pollution, and $PM_{2.5}$, rarely hits the headlines but a staggering fact is that it causes double the annual deaths compared to car accidents in the US alone. There is regulation under EU law to try to curb excessive air pollutants such as NO_2. The EU permits up to 18 hourly breaches of the maximum limits per year per country under Directive 2008/50/EC.

Nitrogen dioxide (NO_2) and nitrogen monoxide (NO) are the product of internal combustion, referred to as NOX. They are poisonous gasses in concentrated quantities. In the 1970s, lead was highlighted as causing health effects, such as brain deformities in newborns. As with CFCs destroying the ozone layer, this was addressed by leaded fuel bans. NOX however has not been addressed. Although partially tackled by the introduction of catalytic converters from the late ‘70s onwards, NOX gasses and $PM_{2.5}$ still make it into our air in massive quantities. NOX gasses combined with particles of organic matter (volatile organic compounds) on a sunny day produces ozone, but not the high-altitude good kind. Excessive ozone exposure is responsible for at least 10,000 premature deaths in the EU per year. Other gasses like Volatile Organic Compounds or VOCs can be produced from trees; these are naturally occurring gasses. Another gas is Sulphur Dioxide (SO_2). Burning coal is a major source of SO_2 and was responsible for the acid rain publicity in the 1980s. Coal-burning has been massively curtailed, and the acid rain story has all but vanished. Destruction of the ozone layer, acid rain, and lead in petrol have all been addressed to differing degrees of success, but NOX and $PM_{2.5}$ remain the most serious air pollutant problem today.

Playing around with my air quality monitor I began to test it in various settings. On a windy day, even in the heart of cities like Luxembourg, Paris and Dublin, the readings were a pleasing 4-8. However, when I was sat at an outside cafe terrace on a cold, windless Dublin day, next to one of the busiest roads in Ireland, the Drumcondra / Swords Road I got scary readings. This major thoroughfare, with four lanes and constant heavy diesel bus traffic, produced a worrying $PM_{2.5}$ level. Less than 2 meters from the roadside, enjoying a cappuccino with my 11-year-old daughter, the device measured above 80 $PM_{2.5}$, the cancer-causing zone! To verify the readings, we moved into the coffee shop, just at the porch door due to Covid restrictions, and the levels immediately dropped to less than 20. Becoming more adventurous and risking my modest investment, I put the device on the road itself. Sitting on the double yellow lines, less than half a meter from the traffic, the device recorded 105um. Again, to verify, I brought the device with me to a nearby park, 200-300 meters from the busy street where it recorded a level of just 10… and that's just $PM_{2.5}$. Microparticles, even more damaging to your health also linger in the air, nanoparticles, or ultra-fines of diameters of $PM_{0.1}$. Coronavirus has a diameter of between 50-200 nanometres or $PM_{0.05}$ to $PM_{0.2}$. Long term proximity to major city roads on windless days is probably more life-threatening than walking across the street blindfolded!

The smaller the particle, the more likely it is to penetrate your various body tissues causing inflammation and cell mutations resulting in cancers and strokes. That SUV driver that obliviously contaminated my morning cycle air in Luxembourg receives more than I do on my bike when he is sitting in traffic. The next time you're sitting in heavy traffic, especially on a colder day, observe that perceived 'steam' rising from the exhaust in front of you. Remember the concoction of chemicals it contains. These particles are smaller than your cars air filter. It's in your car in high toxic concentrations. To prove this, I put the monitor with a piece of Velcro on the dash of my electric car. On a long fast drive to Brussels, it rarely recorded over 3… but when sitting in heavy traffic on the Route d'Arlon in Luxembourg with the vents open and air blowing, it recorded 45! Ironic to be sitting in a zero-emissions car whilst risking my health from my fellow oil-burning commuters. Just like passive smoking, pre indoor smoking bans, more on this in the next chapter.

DIESEL LOVE AFFAIR

The diesel love affair was and still is a toxic relationship.
Europe especially has had a love affair with diesel over the past thirty

years or so. Back in the 1980s and early 1990s, diesel cars were still noisy, heavy, and often unreliable machines. I remember my father's first diesel car, a Ford Granada, in 1990. In the middle of winter, he would turn it over early on a cold morning and I could hear the chug-chug of the engine as it sparked (or glowed rather) into life. With just a few months on the clock, it spewed black smoke when it started. Diesel was not fashionable back then, not sexy, it was reserved for farmers, truckers, and taxi drivers. This all began to change as the price differential between petrol and diesel grew wider and the technology began to improve. The big German manufacturers began major research and development programmes on diesel engine efficiency, noise, and performance in the mid-eighties. The result of this work produced quiet, smoother, and more powerful engines. By the mid-1990s, diesel cars began to display new badges on their rears like Turbo Diesel, Intercooler, HDI and TDI. They began to become sexy and mainstream. I can remember the TV ads where a glass of water was put on a running diesel engine to show just how smooth they had become, or when the new car owner explains his car was diesel to his neighbour's surprise.

Diesel cars are not all bad. They are the most efficient form of internal combustion propulsion. This is thanks to higher air compression ratios and ignition systems compared to gasoline. They are great for heavy equipment and machinery, generators, and locomotion. Lower fuel prices, often combined with road tax incentives for diesel, mixed with great technological advances, plus marketing, ensured a golden age for diesel cars. With better fuel economy, more advanced performance matching petrol, and altered consumer perception, diesel gradually became the favourite of European car buyers. Clean diesel cars seemed to have arrived and no longer were they seen as the heavy, unglamorous workhorses, but rather as the new thoroughbreds. "Diesels are cleaner than petrol cars per kilometre and they last longer." This was the generally held perception of the motoring public for at least twenty-five years. It was, however, all built on a lie.

DIESELGATE

Diesel is dying. The once wonder fuel has been tainted beyond repair, like cigarette smoking. The OEMs, (original equipment manufacturers) were lying, diesel is not clean. Diesel is a killer fuel. Toxic poisonous gas and particular matter are expelled from every diesel engine today. The great advances in engine technology did make them quieter, smoother, and faster, but it masked what was going on. The technological improvements largely eliminated the obvious black soot tailpipe smoke from diesel cars, but the invisible chemical output remained. The extent of these emissions was deliberately hidden from regulators and environmental testing by clever engine management software known as cheat devices. Just like breathing shallow or holding your breath when you were a teenager and didn't want your parents to know you smoked, diesel cars faked their breath tests. This was deliberate, orchestrated and endorsed by the senior management of major OEMs.

Volkswagen was caught in the act and fined €31.3 billion by the EU. In the US, a federal judge ordered VW to pay a $2.8 billion criminal fine for "rigging diesel-powered vehicles to cheat on government emissions tests". Without the 'cheat device', VW diesel cars produced NOx emissions up to 40 times the EPA permissible limit. VW used European grant money meant to reduce car emissions, to develop the cheat device. They did this by programming their cars to detect the patterns of engine revolution and exhaust pressure during emission regulation testing. The result was the cars slowed the revolution speed of the engines and altered the air-fuel mix to produce misleading emissions results. It was wide scale.

Diesel cars produce 15 percent less CO_2 than petrol cars due to their efficiency edge. With this perceived 'cleaner' per mile efficiency, governments reduced road tax to encourage adoption. In Ireland, for example, you could buy a BMW 2.0 litre diesel and pay less than €200 annual tax compared to over €800 for a petrol equivalent. Since Diesel Gate and the growing awareness of our climate crisis, it just seems ridiculous to be receiving a financial incentive to poison our air in the 2020s.

Not just Volkswagen, but several major diesel engine manufacturers were also caught red-handed. This was similar behaviour to the major tobacco companies in the 1950s lying about the effects of smoking. They knew smoking was causing cancer but deliberately pushed misinformation into the public domain to confuse the public and create doubt using orchestrated marketing and media campaigns. As this came to light in the 1990s, the subsequent class action lawsuits cost the tobacco industry billions and forever destroyed the 'it's cool to smoke' public face of tobacco. Advertising bans, health warnings, together with public anger has forever relegated smoking to an outlier, outcast, dirty habit. The Oil industry is now on the brink of becoming a pariah too. VW is now pumping billions of euros into the development of their ID electric car range in the realisation of the electric revolution. Not just Volkswagen, but several major diesel engine manufacturers were also caught red-handed. This was similar behaviour to the major tobacco companies in the 1950's lying about the effects of smoking. As of November 2020, VW has indicated they will commit over €70bn in EV and EV software development and there's a lot more about this later in the book.

AIR MADE DISEASE

Cancers are mostly man-made. Lung cancer, for example, is a direct result of smoking tobacco smoke in over 80% of cases, according to the US Centre for Disease Control. More obscure cancers, such as fast-killing pancreatic cancer, can be linked to a build-up of toxic alien particles in the body, such as Diesel PM. The affected cells try to compensate and react to these foreign bodies by mutation. This mutation goes wrong and writes destructive DNA level code into our cells. This code causes erratic and abnormal cell mutations that spread. We have made great strides in treating the symptoms of this terrible disease but not the cause in most cases.

Air pollution shortens life expectancy, it has made Covid survivability worse in heavily polluted cities. Human activity is the cause. Chemicals such as those released in internal combustion cars, including polycyclic aromatic hydrocarbons (PAHs) cause cancer. Diesel and petrol exhaust gases contain some of the following killer chemicals, classed as group 1-2 carcinogens and group 3, possible carcinogens: carbon monoxide, nitrogen oxides, sulphur dioxide, nitric acid vapour, benzene, chlorine, acetaldehyde, mercury, and arsenic. If you live next to a major road, street or highway, your chances of developing lung cancers increase by at least 10% according to a study by King's College London. Air pollution, particulate matter, and exhaust derivates can kill you. The OECD estimates that a 1µg/m3 increase in $PM_{2.5}$ concentration costs the EU almost 1% of GDP, equivalent to €180 billion. In the US, air pollution is responsible for approximately 200,000 deaths each year and creates a 4% hit on GDP, equivalent to around $790bn per year. Compare this to the investment made in electric charging infrastructure, representing just over $500m according to the US-based Alliance for Transport Electrification, just 0.6% of the air pollution cost.

Multiple studies have determined a direct link between air quality and health. From causing breast and lung cancers to damaging the liver and kidneys; essentially every part of the body can be affected by $PM_{2.5}$ or lower. Nanoparticles become absorbed through the nasal, throat and lung membranes to penetrate the bloodstream. They can penetrate and inflame the walls of arteries, the heart and brain tissue which can ultimately result in heart attacks, and strokes.

Diesel fumes impact around 400 different genes, with prolonged exposures causing mutations and cancers. Studies from Hong Kong linked PM concentration to increased breast cancer risk. Long term air pollutant exposure has been linked to Parkinson's disease in one Korean study and another UK study, increased risk of dementia. European studies linked traffic-caused air pollution with cognitive development in young school children.

ICE cars should have health warning stickers on them like cigarettes have "WARNING! MAY CAUSE CANCER" or "THIS CAR REDUCES YOUR LIFE SPAN" etc. Of course, this is unlikely to happen, but why not? Big oil and vested interests are today spreading misinformation just like tobacco companies did in the 1960s through to the 1990s. They are also running superficial 'greenwashing' programmes to fool the public. The scale of deception and misinformation through lobbying and marketing is incredible as I will elaborate on later.

WHO'S TO BLAME?

The amount of CO_2 in the atmosphere reached record levels in 2020, 417 parts per million in May. The last time atmospheric CO_2 was at these levels was during the Pliocene era, around 4 million years ago, when global temperatures were 2-4 Celsius warmer and the sea was 10-25 metres (33-82 feet) higher than today. It takes a long time to melt kilometres thick ice sheets, but they are melting, and the melt will match the CO_2 of today with the sea levels of the Pliocene area. With more than half the world's population living in coastal regions, even a 10-meter sea rise would be catastrophic on a biblical scale. If you think this won't affect you or your loved ones in your lifetime then think again. 2010 to 2020 was the hottest decade on record. 2020 was 1.2c hotter than the average for the entire 19th century. This heating has had a dramatic impact on the Polar Ice sheets. Antarctica has been frozen for 30 million years and the Earth's population has more than doubled since man walked on the moon in 1969. Not only has the Arctic ice cap witnessed the largest shrinking ever in 2020, across the northern hemisphere, permafrost, but the frozen ground is also warming alarmingly. Siberia reached a temperature of 38C (100F) in the summer of 2020. Melting permafrost releases another climate killer - methane, in addition to even more CO_2. This combo effect of human-made greenhouse gasses combined with released permafrost gasses can cause the infamous runaway effect, like trying to put out a fire with jet fuel.

Who's to blame? The answer is, all of us, however, some countries are more guilty than others. China, India, the USA and Russia account for over half the CO_2 and other greenhouse gas emissions in the world. Combined, they produce over 20mt (megatons) of CO_2 Emissions. A megaton is one billion kilograms. For example, 10 kilograms is the average amount of gas contained in your barbecue gas cylinder. So, these countries produce the equivalent of 2 billion gas cylinders of CO_2 per year, almost a quarter of a million canisters every hour.

The worldwide total in 2017 was 37mt CO_2 according to the UNFCCC (United Nations Framework Convention on Climate Change), making these four nations responsible for 55% of the global total alone.

On a per-capita basis, we see a big divergence, with the USA producing 15.7 tonnes of CO_2 per year per person, whilst Indians produce just 1.8 tonnes of CO_2 per year or just 11% of an equivalent American. The Chinese produce around 50% of the average American at 7.7t. With the rapid rise in Chinese and Indian living standards, if they were to reach the American levels of CO_2 per person, then humans would emit a staggering 73mt of CO_2, double today's level even with no population increase. It's unfair to single out any one country, there are hot spots of air pollution in Western nations on par with cities in India and China. Krakow, Poland (close to the geographical heart of the EU) is often a top ten polluter on the Air Quality Index, due in part to its adjacent coal power plants and unclean industries. When taking a global view of air pollution, it's important to remember it's all the same air at the end of the day. Like banning smoking in just one part of a large, closed room, it still affects us all.

Air pollution does not respect national borders, looking on a planet-scale zoom out of India on the IQair.com website glaringly demonstrates this. You will see a massive yellow-orange and red cloud (classed as unhealthy air) pouring out from India into the Indian Ocean towards the Middle East and as far as the eastern Mediterranean Sea. Indian air pollution is carried extreme distances by fast-moving high-altitude winds. We all breathe global air pollution regardless of the source.

The good news is neither China nor India is blind to their impact. China is making an enormous large-scale push towards mobility electrification and clean energy production. India is building an imperative to transition to Electric Vehicles and cleaner sources of power production. As of 2020, India is a net importer of polluting fossil fuels equivalent to $130bn annually. But days of extremely high heat and extreme air pollution are both increasing worldwide. For example, New Delhi experienced a week of the worst air pollution in human history in November 2017. The air quality was so bad that the airport had to shut as planes could not land. A study at Texas A&M University concluded that the likely frequency of days with both extreme heat and pollution simultaneously occurring could massively increase by 2050. Global CO_2 emissions have more than doubled since the 1970 level of around 15mt per year. The culprits for this increase are mostly China, India and nearby Asian countries. In the USA and EU CO_2 levels have been mostly static or declining with the advent of clean air acts in the 1970s prohibiting open burning coal and other solid fuels. However, the rate of decline would have been much greater than for the massive increase in new vehicles. The growth has been staggering since World War Two when CO_2 emissions were just 12% of what they are today. In the United States, transportation is the number one contributor to emissions.

Pointing the finger at the worst offenders of today leads to a battle about who was responsible for historic pollution. In every climate debate, the developing economies cite that the developed, western economies, polluted our skies for centuries and that their 'Carbon Legacy' is far more than their own. The developing countries demand the right to enjoy a standard of living on par with the West, and so, they should. The issue is how this can be done with new green energy and clean mobility solutions to prevent them from becoming dependent on fossil fuels.

Historically, the USA is the worst offender. The US is responsible for 25% of total historical emissions, the EU 22% since the industrial revolution. 5 tons of CO_2 emissions per person is the global average. Straight-line averages are however misleading. Australia has the highest per-person output at 17 tons. The US is next worst at 16, 10 for Germans. To put Western emissions into perspective with the developing world; an average American produces more CO_2 in three days than a Nigerian does in one year. Ironically, the countries emitting the least stand to lose the most. Climate change drives typhoons, hurricanes, rising sea levels and ecosystem destruction that impact these poorer geographies the worst. The stark fact is that the world's richest half of the population, of 3.6 billion people, produce 86% of global emissions. As these poorer countries develop and demand the lifestyles of the West, low carbon energy development is vital. Just like many African nations skipped the installation of a twentieth-century wired telephone network with the advent of mobile phones, they need to skip the fossil fuel era also. The planet simply can't take it otherwise. I'll explain this at the end of the book.

The scale of man-made air pollution is staggering. Internal combustion engines alone emit 22 million tons of CO_2 into the atmosphere each day or 8 billion tons each year. To put that into context, the 2010 eruption of the Icelandic volcano Eyjafjallajökull, emitted just 150,000 to 300,000 tons CO_2 per day into the atmosphere for 3 months. To avoid possible jet engine damage from micro volcanic particles, airlines diverted flights around the world. I flew transatlantic in May that year and the pilot diverted the plane to the north of Iceland. I was fortunate to see the volcano from the air and it was truly awe-inspiring and terrifying what our world can do. It was dramatic but represents less than one-tenth of a per cent of the annual oil burns from internal combustion.

There are estimated to be 1.4 billion cars in daily use on the planet. Electric cars need electricity, and yes this is still mostly generated by fossil fuel burning power plants (more about this in a later chapter) but at least it is centralised in one location. Not good of course for the people that live close by, but far better than 1.4 billion mini power stations driving all over the world. In later chapters, I will explain the energy required and sources to fully switch all mobility to electric.

COVID19 WAKE UP

In António Guterres 2020 speech he also said, "Nature always strikes back". Covid-19 is largely man-made, created by a combination of climate change and biodiversity destruction. We released an animal virus into the human health chain by directly invading natural spaces, poisoning the air, and mixing different ecosystems that were never meant to be mixed. The result was the global Pandemic that started in 2020. The first in modern history and not the last. The UN has said that a pandemic such as Covid-19 will happen again. That's almost a certainty. Our solution has been to destroy economies with lockdowns, pump billions into drug and vaccine research. As with the great world wars and the quest to put a man on the moon, humans under pressure, united, can achieve anything, and fast. The development of the Covid vaccines during the summer of 2020 was achieved at a lightning pace. What would normally take at least 3 years was done in less than six months. Companies and organisations such as Moderna, BioNTech and Pfizer developed, tested, mass-produced and distributed millions of vaccines around the world in incredible record time. Partly driven by profit and by our survival as a world economy, we as a species achieved something that would have been considered impossible just one year before. Imagine if this level of focus, purpose and investment was put into tackling our air and climate crises? The problem is, we are irrational apes. Designed to deal with immediate threats with our big brain solutions. Despite all the plans and strategies, we are, in the main, very short-term thinkers with short term memories. Covid will eventually fade into another major historical event, talked about for a few years, then to be become an 'I remember when' to our kids and grandkids.

The good news is that Covid has awoken many to the air they breathe. Who would have thought just one-year later millions would be wearing masks all over the world? Covid, combined with planned city/country bans on internal combustion car sales and a deluge of new electric cars coming onto the market has lit the spark for an Electric Revolution. In 2020 sales of new electric cars surpassed diesel cars in many European countries. Over half a million BEVs -Battery Electric Vehicles- were sold in Europe in 2020. Electric car sales grew by over 70% in 2019 whilst sales of diesel cars dropped by as much as 80% in some markets such as the UK. The tipping point has tipped but electric cars represent still only a tiny fraction (4.8 million according to the International Energy Agency) at 0.3% of the more than 1.4 billion polluting cars on Earth. So, what can be done to clean our air?

WHAT CAN BE DONE?

Covid-19 has devastated the global economy, resulting in governments having to inject trillions to prop up economies. Stock markets in the US had been driven to all-time highs, not by the reality of the dire impact of Covid but by the artificial support of ultra-low interest rates and state supports. This cannot continue indefinitely. Without these artificial supports a great recession, on par or worse with 2008/2010, will happen. The pandemic is a critical and awakening moment to focus recovery efforts on clean energy, green technologies and a new world. The opportunity is there. The EU, for example. has set a target of 2050 to reduce transport sector emissions by 95%, although this is nowhere near enough and frankly an outdated objective.

Time is not on our side. Projected CO_2 emission targets necessary to limit global warming to below 2C above pre-industrial levels require radical action. Right now, projected emissions are way above what is necessary to effect change. Simulation projections require a continuous drop in CO_2 emissions equivalent to around 10 metric gigatons per year to offset population growth and reduce overall emissions. To put that into context, emissions have grown from around 35 metric gigatons per year in 1990 to around 50 in 2020. The target to limit global warming to 1.5C requires zero CO_2 emissions by 2090. Current projection models (even taking Covid into account) predicts an actual increase to 60 gigatons by 2050 before beginning to drop. This is way too late and will result in global average temperature increases above 2C, resulting in catastrophic climate damage and the predicted 'century of hell' in the 2100s. This is not a pleasant legacy our children and grandchildren will inherit.

Whilst transport and related energy sectors are only part of the problem, it's a growing part with increasing population growth and more cars hitting the roads every year. Electric cars combined with clean, renewable energy production are a vital part of the solution. The technology now exists but the will is still missing. Despite the Covid crisis, some countries are using the pandemic to direct incentives to the wrong solutions. Poland, for example, is spending $25M acquiring unwanted coal to relieve its industry. Canada is investing over $1bn in Alberta to establish a new oil pipeline. Australia is developing new coal mines in Queensland. These are the wrong solutions at the wrong time. Sustainable projects and the full electrification of mobility is the future. In his State of the Planet speech, Guterres stated that the 'central objective' was to build a global coalition to achieve net-zero. Immediate actions are necessary. We must make the change and now.

It takes bold decisions, a leap of faith and courage to make the shift. Humans are the most adaptable animal on the planet, we quickly become accustomed to the new norm. We can adapt, adopt, and improve our world quickly to new challenges, we just need to do it. Having lived and studied the electric revolution intimately for the past five years and echoing some of Mr. Guterres pleas, here is my top 10 to accelerate electric mobility and move to a world with clean air.

- Put a price on carbon
- Ban new petrol & diesel car sales globally by 2030
- Phase out all fossil fuel subsidies
- Change the tax system from income to carbon
- Put a health warning on cars and petrol stations
- Rapidly rollout simple EV charging infrastructure
- Mobility sharing & pooling incentives
- Immediate city centre internal combustion-free zones
- Incentivise clean energy production globally
- Implement a worldwide explosion in solar energy

The election of Joe Biden and the US re-joining of the Paris Climate Accord is a step in the right direction after the neglectful, damaging Trump years. John Kerry was nominated as the Climate Tsar, already demanding China reach net zero by 2060. The Apollo astronauts had what they called the 'right stuff', the question is do we have the right stuff to put selfish interest beyond our planet's survival? Before I go into how anything can change, we need to tell the story of BIG OIL and how the fossil fuel industry is covertly fighting back.

CHAPTER 2

BIG OIL

THE HYDROCARBON GAME

Nothing good ever comes to mind when the words 'BIG OIL' are spoken. Oil was one of our greatest discoveries, propelling incredible human advances but could also be possibly our biggest mistake. I started collecting the National Geographic magazine when I was 12 years old, in 1984. The spectacular cover images for which the magazine is famed captivated me. One cover image struck and stayed with me, the January 1990 cover of the Exxon Valdez disaster. The catastrophic oil tanker spill in pristine Alaskan waters. Images of rare, protected birds covered in oil on beaches, slowly suffocating as they tried in vain to flap their wings and escape into the crystal blue Alaskan sky. It was sheer horror and at the time sparked massive public outcry.

Exxon's giant oil tanker spilled 11 million gallons of crude oil into Prince William Sound on March 25th, 1989. The oil slick covered 1,300 miles of coastline and wiped-out hundreds of thousands of birds, fish, seals, and whales. The captain, Joseph Hazelwood, had been drinking and allowed a rookie, unlicensed third mate to steer the ship. Hazelwood was fined just $50,000 and ordered to perform 1,000 hours of community service. The repair work took years and cost Exxon over $2 billion in clean-up costs. The image of oil companies was irreparably damaged, and new public scrutiny came over their global operations. This resulted in the US 1990 Oil Pollution Act - increasing penalties and sanctions on future oil spills. The damage in Alaska was tragic wildlife and ecological disaster but just a tiny blip compared to the global climate catastrophe currently unfolding. Almost entirely due to our insatiable appetite for oil.

Memories fade and what was once big news, attracting worldwide condemnation was soon forgotten to history. That is until it happened again, on the 20th of April 2010. The Deepwater Horizon oil platform exploded, tragically killing 11 BP employees. The Gulf of Mexico was devastated. Once again, the images of wildlife washing up on gulf state beaches filled the media evoking horror and outrage. The cost to clean up this disaster was $65 billion, including fines of $4.5bn. There have been bigger oil spills, like the 80-million-gallon Russian pipeline breach into the Kolva River in 1984, but Deepwater's proximity to mainland USA ensured greater media attention and public anger. Once again memories fade and people forget. The oil companies are relentless, going further and deeper than ever before in the search for black gold. This is a complex, dirty business and when, you stand back and look at it, insanely dangerous and unnecessary - an outdated legacy of the 20th century. We need to ditch the oil habit, fast.

Humans have used petroleum (or oil) for thousands of years to make medicines, illumination and weapons. Evidence of Chinese oil transportation via bamboo has been discovered as far back as 600 B.C. There has been a petroleum industry of sorts for millennia. However, this usage was infinitesimal compared to the surge since the Industrial Revolution. People in the business call it the Hydrocarbon industry. It sounds fancy, scientific, almost elitist. It means drilling, production, refining and burning black oil. Serious production began in the US in the 1860s following the discovery of oil and subsequent drilling at Oil Creek in Pennsylvania in 1859 by Edwin Drake. Oil production started slow, at first, and then quickly gained momentum and soon went exponential. Growing global oil appetite was fuelled by the advent of the internal combustion engine, the mass-produced Ford Model T and an explosion of new automobile companies across the globe. 100 years later, in 1959, oil production had developed worldwide in thousands of oil fields producing 20 million barrels of oil per day. This has since grown to 95 million barrels produced per day in 2021.

Humans have developed an addiction to oil. Entire economies have been built on its production and refinement. There are, and have been, oil supply crises and cartels influencing the price of oil for over a century - causing geopolitical instability, economic recessions and wars. The burning of oil and oil-related products is the primary cause of climate endangerment, global warming and air pollution. The Intergovernmental Panel on Climate Change (IPCC) concludes that the fossil fuel industry is responsible for 89% of CO_2 emissions and associated global warming.

History won't look kindly on the oil age.

Refined, it produces a great degree of energy at a relatively low cost. This energy has powered our cars, planes, ships, trains, homes and factories. It powers our electricity stations and produces tires and plastics. Oil is prevalent in millions of everyday products. You can't completely put the oil industry down. It gave humanity the kickstart to exponentially develop. Oil provided the spark that ignited the fire to propel humanity, in the past 150 years, beyond even the wildest dreams of previous generations. However, like the end of the stone, iron and bronze ages the oil age is coming to an end. It was brief, relative to human history, incredibly beneficial to human development, however, catastrophic to the health of our planet. In just 150 years we have near destroyed our climate and increased global temperatures to the point where our very existence on this rock is in real jeopardy.

The Earth is a giant rock floating in a cold, vicious, inhospitable emptiness. Life, complex life, has no business existing on it, but for luck. Our rock is in just the right position to our sun to allow liquid water, 'the Goldilocks zone'. Liquid water is the universal molecule making all life possible. A little bit closer to the sun and we are Venus, with iron crushing atmospheric pressure and a toxic-life-eliminating climate. A little further away and we are Mars, an icy cold, desolated wasteland. We are lucky, extremely lucky, to have a moon of sufficient gravitational pull to mix up the primordial oceans with tides and create a life producing smoothie mix of basic elements, complex chemicals and enzymes that became the first lifeforms. We are lucky our planet's massive collision with an early solar system object created the moon and left us with a limp that allowed for the tremendous diversity of life. This limp is a 23-degree tilt to our axis around the sun, giving us the seasons. Our relatively short day (just 24 hours compared to almost four Earth months for a day on Venus) means the Earth, as it spins, never stays in the sun's glare for too long to overheat or in darkness to freeze. The seasons accelerated the complexity of life as organisms adapted to survive in different light and climate conditions. All this created the trees, plants, fish, birds, the dinosaurs and us. As a result, our big round rock is thinly coated with a layer of life. Like fine moss on a boulder. This layer of life exists under the blanket of an ultra-thin atmospheric wrap, just 100 km high.

Life shouldn't exist at all. Even with our atmosphere conducive to life, even with the right position to the sun, even with the moon to churn up our seas and even with our seasons, space would never normally have allowed us to exist. Again, we are lucky. The centre of the Earth has an iron molten core that spins creating a magnetic effect. This gives us north and south magnetic poles but more vitally our shield, our magnetic field. This shield also called the Geomagnetic Field, protects us from solar radiation. The emergence of our solar shield corresponds to the first basic life forms on earth, around 3.5 billion years ago. The sun spits out terrifying radiation and solar particles that would wipe out life instantly without our shield. The aurora borealis is a beautiful display of this shield pushing back these killer particles - a fortuitous phenomenon necessary for survival. With all this luck we have the perfect habitat for abundant life, complex, beautiful and sometimes strange. However, our habitat is extremely finely balanced. Our existence is within a tiny band of temperature range and chemical composition. Zero to one hundred degrees Celsius, ice to steam, is a tiny range on the universal temperature scale. Imagine a straight road one thousand miles long. We can survive only between miles 500 and 501. That's it. Mile 499 and we freeze, mile 502 and we evaporate.

The Earth is self-healing. Over millennia, ice ages caused by catastrophic volcanic eruptions and Earth's orbital wobble, plus heat waves, earthquakes and even cataclysmic asteroid strikes were self-repaired. The asteroid that caused the dinosaur extinction was estimated to have killed off 99% of all life and still, the Earth healed and recovered. This delicate balance of nature ebbs and flows through the ages, sometimes it's been a little warmer, sometimes a little colder but the change has been gradual over hundreds of millions of years. Then we came along. We have interfered with the natural system, and not just in a superficial way but at a fundamental level - because of oil.

BURN IT ALL

I regularly fly from Dublin to Luxembourg. There are no electric plane options yet. As you approach the coast of the Netherlands and look out the window, you will see the massive port of Rotterdam. To the south of the mighty mouth of the Rhine River, there are multiple oil refineries located at this nautical gateway to Europe. BP operate one of the largest in Europe here. Even from a height of 10 kilometres, you will easily spot a massive burning flare. It's especially visible at night. When I first saw it, I was impressed! What is it? It is the burn-off or gas flaring, as it's called in the oil industry, of excess gas from the refining process. Burning equates to heat, to energy, however, it's thrown away. It is not economical to recover these gases and use them for other uses - oil is easier. Instead, it's burned in a gigantic flame into the atmosphere. Gas flaring is just one relatively small heat and CO_2 source produced during oil production. This happens on a global scale in over 700 oil refineries. Gas flaring produces just 1% of the global CO_2 emissions, however, this is a total waste of potential energy.

In 2019 I flew at night, from Dublin to Dubai. I had never taken this trip before. I can never sleep on flights and obsessively watch the little aircraft's progress on the map screen. I could see we were over northern Iraq / Iran. When I looked out of the window the sky below was cloudy and glowing red-orange for as far as I could see into the vast horizon. I thought we were over a large city, however, looking again at the map there was nothing of note on the ground below. Around 30 minutes passed, and the sky was still glowing orange until a brief gap in the cloud cover revealed the culprit below. Hundreds, possibly thousands of ignited gas flares. This is oil country. The region I had been viewing from my window seat was located around the Kirkuk oil fields, in north-eastern Iraq. These vast oil fields straddle the borders of Iran / Iraq and are estimated to contain 16 billion barrels of oil reserves, one of the richest reserves in the world and the site of deadly past conflicts. Regardless of the political assertions, weapons of mass destruction claims or terrorist links, past wars here were because of oil. Compared to this region the Rotterdam gas flare seemed like a single match next to a nuclear explosion. The waste is staggering. Not only are we extracting the accumulated past and burning it in our engines, but we also so disregard our precious atmosphere we burn off the less profitable excess gasses in vast quantities. The value of this gas loss is estimated to be worth over $19 billion per year, according to a 2019 study by GlobalData. This energy is simply discarded into our atmosphere in total contempt for our climate.

All this burning has not come without consequence. Oil could be our biggest mistake. The result of over 150 years of burning the fossils from the past has been catastrophic for the delicate balance on earth. We have destroyed our planet's atmosphere and biodiversity. Before oil, Earth experienced a relatively calm climatic period. In the past 2.6 million years, known as the Quaternary period, and apart from ice ages cause by periodic wobbles in earth's rotation around the sun, it's been bliss. These last 12,000 years have been particularly tranquil, known as the Holocene epoch. During this period predictable seasons allowed life to flourish and adapt. This life and biodiversity in turn help to balance the atmosphere by absorbing CO_2, methane and other greenhouse gasses. This rhythmic absorption and release, self-cools the planet.

Other elements like the vast polar ice caps help reflect a large proportion of the sun's energy into space, further cooling our world. It's not just the emissions of cars but the extraction process of oil itself from the ground that produces unsustainable levels of CO_2. The movement of crude around the world (powered by diesel tankers) is hugely detrimental to marine life. There are estimated to be over 800 very large crude carriers (VLCC^s) in service worldwide, each carry around 2 million barrels of oil. These giant hulks, with a full belly of black gold, produce over 100 million tons of CO_2 per year alone as they traverse the world's seas. The oil then needs to be refined before being again shipped by pipelines, rail and road tankers to your gas station. Oil is moved by burning more oil! Scientists have concluded the Holocene is over. According to the International Commission on Stratigraphy, we are now in the Anthropocene period. Human activity ended the Holocene epoch, abruptly.

TOO MANY OF US

The climatic damage done by oil has been compounded by population growth. The human population has more than doubled since man walked on the moon, just fifty years ago. The Earth could sustain this number and more but not the way we currently run things. Our pace of progress and the methods that got us this far will get us little further. Continuing to burn everything, to consume everything and eat everything at the rate of the past is not sustainable into the future. With further population growth and finite natural resources, disaster looms. The climate's warming will go into catastrophic autopilot mode if we do not make radical changes. With melting ice caps, the sun's formerly reflected energy is being sent into the oceans. The oceans have already been a temporary buffer, by absorbing excess heat. Unfortunately, the ocean's ability to absorb more heat and continue to regulate the temperature is almost over. The energy from the heat in the warmer oceans has reached capacity. This energy is released in violent storms, unpredictable weather events and catastrophes. As the Earth's permafrost regions begin to melt, they release long-stored methane gas. Methane is far more powerful than carbon dioxide in trapping in atmospheric heat; 84 times worse. The combined effect of methane release, disappearing ice caps and further fossil fuel burning will trigger a tipping point. When our natural global air conditioning breaks down, the planet warms uncontrollably. It will be too late. As with past extinction events, life may recover in the long run, but we will not.

As Earth evolved over millennia and life came and went, the remnants of that life, and the carbon it contained, were 'stored' away - forever - it's not supposed to come back! Geology is the study of time. When the dinosaurs died off, together with millions of other plants and animals, the carbon in their bodies broke down into compounds. Over time the layers of dust and ash built up to form rock. The pressure of gravity gradually crushed these compounds into liquid oil. It's deep underground in most cases, where it was supposed to stay, forever. That is until a small hairless ape took a liking to it and decided to bring it back to the surface! The cumulative effect of deforestation, destroying the lungs of the world, combined with the burning of fossil fuels has increased the carbon dioxide content in the atmosphere to dangerous levels. Dramatic climatic events have caused at least five mass extinction events in the history of life on Earth. From volcanic to asteroid strike causes, all have been associated with a rise in greenhouse gases, especially CO_2. The collapse of ecosystems and the increase in atmospheric CO_2 during these events happened relatively gradually. Humans have, however, increased the CO_2 level by more than the equivalent of one million years of volcanic eruptions in just two hundred years. Instead of a slow steady change over millions of years, our impact has been immediate. A continued rise in CO_2 levels at the same pace will lead to a planet in which humans cannot live in less than one hundred years or less. The Earth can simply not recycle itself fast enough to cope with our scale of impact. Oil has simply got to go. Now.

YOUR DOCTOR RECOMMENDS CAMELS

For over 60 years, Big Oil has known about fossil fuel induced climate change. Like the tobacco industry from the 1960s to the 2000s, big oil lied about its climate impact. Worse still they conducted, and continue to conduct, a deliberate policy of misinformation to confuse the public. The insidious similarities between the tobacco and Oil industries range from negative campaigning, false advertising and lobbying to intentionally discrediting climate change, EVs and clean energy.

Edward Teller, one of the leading scientists on the atomic bomb Manhattan Project, was invited to speak at the 100th anniversary of the American oil industry, in 1959. At the event, organised by the American Petroleum Institute (API), he told the hundreds of gathered oil executives that we needed to look at alternative fuel sources. He then explained that CO_2 in the atmosphere causes a greenhouse effect and continuing to burn oil would melt the ice caps and put cities, like New York, underwater. He quite accurately predicted the rise in CO_2 levels and associated temperature increases we witness today. One of the 300 people in that room was the Chairman of the API, Robert Dunlop. He later testified to a Congressional committee on the topic, "Tomorrow's car: electric or gasoline-powered?" He told the committee, "We in the petroleum industry are convinced that by the time a practical electric car can be mass-produced and marketed, it will not enjoy any meaningful advantage from an air pollution standpoint" The campaign of confusion directed by the oil industries 'merchants of doubt' was well underway.

In 1965, then president of the API, Frank Ikard warned 'time is running out in a speech on climate change. He was referring to a report commissioned on climate change in which he said, "The substance of the report is that there is still time to save the world's peoples from the catastrophic consequence of pollution, however, time is running out." His speech was published in Nature Magazine, one of the most damning statements reads as follows, "One of the most important predictions of the report is that carbon dioxide is being added to the Earth's atmosphere by the burning of coal, oil, and natural gas at such a rate that by the year 2000 the heat balance will be so modified as possibly to cause marked changes in climate beyond local or even national efforts," In 1968 the API hired scientists to investigate global warming. They were again warned that significant temperature changes could occur by the year 2000. Exxon produced a report in 1978 on limiting CO_2 - the report suggested leaving fossil fuel reserves unexploited - the report concluded that only about 20% of remaining fossil fuel could be exploited without doubling CO_2 in the atmosphere.

The predictions were very accurate, indicating 400 parts per million of a CO_2 increase would produce a 1.5-degree Celsius average global temperature increase.

James Black, working under Exxon's Products Research Division, wrote an internal briefing paper called "The Greenhouse Effect" - warning that human-caused emissions could raise global temperatures and result in serious consequences. "Present thinking holds that man has a time window of five to ten years before the need for hard decisions regarding changes in energy strategies might become critical". In response to the report, Henry Shaw from Exxon wrote a memo to senior execs on how they could influence any possible legislation on environmental controls. A coordinated campaign began industry-wide, leading in part to the establishment of the API's 1979 CO_2 and Climate task force. In 1988, Shell produced an internal report predicting the impact of climate change, including flooding, storms and loss of biodiversity. Big Oil's response was to try to systematically discredit climate science.

Big Oil set up the 'Global Climate Coalition' to spread misinformation and try to disprove the link between fossil fuels and global warming.

In 1990 the Intergovernmental Panel on Climate Change confirmed that increased carbon dioxide would cause global warming. In 1993 US president, Bill Clinton proposed a carbon tax - only to be defeated by the API and the Global Climate Coalition. In 1997, Lee Raymond, CEO & Chairman of Exxon, gave a speech at the World Petroleum Congress in which he spread doubt on climate change and the actions necessary to avert disaster. He stated that "the case for so-called global warming is far from airtight" in challenging the Kyoto protocol whilst acknowledging that this is the greatest threat to the oil industry. The Global Climate Coalition's lobbying and misinformation made sure the Kyoto protocol was not adopted and defeated with the 2001 withdrawal. During his ill-fated presidency, Donald Trump cited the same Big Oil scientists when he broke from the Paris Climate Accord. The oil industry knew their actions were causing climate change for over 60 years. They knew temperature rises would be inevitable and catastrophic in the longer term. Concealing what they knew and running a campaign of deception is the greatest, and most tragic, human cover-up since cancer and the tobacco industry.

[1] **SOURCE:** Inside Climate News.

In the depths of the 2020 US election, with significant democratic campaign emphasis on the environment, Minnesota's Attorney General, Keith Ellison, filed to sue ExxonMobil and the API. In his case, he describes "a campaign of deception" orchestrated by the defendants and "executed with disturbing success." Like the tobacco industry decades earlier, Big Oil has a fight on its hands. This is not going to simply go away. Whilst it is easy to point a finger at the hydrocarbon industry, they are not the only ones. Car manufacturers and the mobility industry have all been encouraging consumer demand for cheap easy fuel. Most of us have been largely oblivious to the damage our engines are doing, or at worst, been indifferent to climate change.

Like with the tobacco scandal, industry insiders did speak out on occasion, only to be castigated or to resign in moral disagreement. One such individual was scientist, Marty Hoffert, who worked for Exxon. His climate models predicted climate change and he shared his warnings with his bosses. His superiors acknowledged his finding however he became increasingly frustrated when he witnessed that the public announcements by Exxon contradicted their internal awareness. Hoffert resigned stating "What they did was immoral". They spread doubt about the dangers of climate change when their own researchers confirmed how serious a threat it was."

It wasn't just the fossil fuel industry, but industries dependent on it that began to come together to coordinate their propaganda. Edison Electric Institute (an electrical products industry representative body) launched a campaign called the Information Council for the Environment (ICE) in 1991 – its purpose was aimed to "Reposition global warming as theory (not fact)". Some details of the campaign were leaked to the New York Times. Like the tobacco industry implying smoking was good for you in the 1950s and 60s, this campaign launched false advertising to instil public doubt. Philip Morris cited 'medical authorities' have scientifically proved their brand is less irritating to the throat in one 1950s smoking campaign. In this century, Facebook detected 51 climate disinformation ads in 2020, viewed by 8 million users. For example, one Facebook ad stated, "Make no doubt about it: the hysteria over climate change is to sell you Big Government control." and "Fossil fuels are not an existential threat". Another stated its "mission is to identify, educate, train, and organise students to promote the principles of freedom, free markets and limited government". It was headlined: "Conservatives Are Pro-Science, While Leftists Are Pro-Panic! Climate Change Is A HOAX! #ThinkForYourself #EarthDay."

Cigarette companies had ads with babies on the cover, ads claiming 20,670 doctors recommend Lucky's Cigarettes or "more doctors smoke Camels". These ads were pushed on the public when the tobacco industry knew, without a doubt, their products caused cancer. The gigantic class-action lawsuits that followed in the 1990s onwards cost the tobacco industry billions in settlements like the 'Tobacco Master Settlement Agreement'. It is strikingly similar, the tactics used by the tobacco industry and the oil industry, in hiding the truth and seeding doubt in the public's mind through false advertising, political lobbying and contrarian science. Although the public is fully aware of the cancer risks, millions still choose to smoke. Passive smoking has also been responsible for the deaths of millions. This is the unfair consequence of a smoker's choice to smoke in the vicinity of the previously unwitting victim.

Climate change caused by fossil fuels is still today not broadly accepted. Mass doubt remains today. Great strides have been made to validate the science and educate the public. However, the climate debate continues and just like smokers today, many will be indifferent to the planet-wide danger caused by their actions. Making a climate positive choice to drive an electric car, even when they are more cost-effective and more entertaining, will still be resisted by many. Choosing to smoke, in the now limited public locations, largely just detrimentally affects the smoker. Choosing to continue burning fossil fuels, however, affects us all.

OIL'S CRISIS

In May 2021 a Dutch court ordered Shell Oil to limit its carbon emissions by 45% by 2030, citing that Shell is in part responsible for global warming. This monumental victory for our climate is a wake-up call for the oil industry. As our dependence on oil wanes, oil nations, for years, propped up by black gold, are likely to suffer political instability. Big Oil has an impending crisis on its hands.

The power of Big Oil is on the wane. Almost as quickly as the industry grew, it is now under threat of survival from multiple fronts. The shock impact of the Covid pandemic, in April 2020, reduced the price of a barrel of crude oil to below zero i.e., they paid the buyer to take their oil. In February 2021, BP declared a 96 per cent drop in fourth-quarter profits attributable to the pandemic. The oil machine is delicately balanced with pricing, supplies, storage and demand finely intertwined. In a changing world, the survivors are generalists, not specialists - the oil industry is highly specialised. An attack on the oil industry is very much underway and on multiple fronts: class-action lawsuits over their climate coverups, bans on new exploration, curtailments of fracking, new oil pipelines, intergovernmental climate treaties and now the exponential shift to electric mobility.

These external threats are all combined to make the next decade a decade of trouble, strife and decline for the oil industry. A look at the market capitalisation and share prices for the top 10 biggest oil producers shows at best a plateau, and more often a steady decline for the past five years. The heydays of exploration, growth and riches are long over. This is an industry in terminal decline and crisis. New regulations and governmental roadblocks are on the way. President Biden immediately blocked the controversial Keystone Oil Pipeline, originally blocked by President Obama, then unblocked by Trump. This is another example of the climate positive focused policies of the Biden administration. For the next four years the weight of scrutiny on the oil industry, combined with the electric car shift will mean that even if the next president is pro-oil, it will be too late for the industry. There is no going back now, the momentum against the fossil fuel industry is gaining pace. They know it and are taking steps in increasingly more desperate measures to sell assets, divest and shift their businesses. This did not start with the election of President Biden. Oil companies have known for years that their long-term survival is in question. Yes, there will be demand for oil for decades to come but the Covid pandemic gave them a taste of the future. Low demand, excess supply, dwindling barrel prices and overweight organisations are not fit for purpose in the new green world.

Before the pandemic oil companies had already been under pressure as 'easy oil' reserves began to dry up. Explorations have become more daring, deeper, more remote and expensive. In the 1970s America reached 'peak oil production' of around 10 million barrels per day. Access to easy oil, like the vast Texan reserves, surpassed peak output and began to decline. America knew its oil supply future lay elsewhere. Alliances with the Kingdom of Saudi Arabia for example, where a relatively simple drill in the sand produced oil, became core policy. The geopolitical tensions, Gulf Wars and to a degree, terrorism, can all be linked to the US protecting its long-term oil supply. For three decades oil production in the US diminished and dependence on external sources increased. By the mid-2000s this began to change. The innovation of extracting oil from substrate and tar oil sands began to increase US production. Known as Hydraulic fracturing or 'Fracking' the US became a net exporter of oil in 2011. By 2020 the US was producing over 12 million barrels per day, far surpassing the peak oil of the 1970s. Fracking is a dirty and environmentally costly endeavour. Chemically laced water is pumped at high pressure into ground rock, shale, soil and sands where pockets of oil have accumulated in fissures. The pressure forces gas and oil to the surface (through the natural water table via many boreholes). Compared to 'normal drilling' it's expensive, however as oil prices rose and fracking technology improved, it became economically viable. An explosion of Fracking ensued in the 2000s. As recently as 2015, an industry group lobbied for the UK to start large scale Fracking to ensure industry competitiveness. The lobbying was funded by a task force established by the fossil fuel industry – 'The Task Force on Shale and Gas'. However, fortunately for the environment, in 2019 Fracking had been halted in England as the government stated, "it would not agree to any future fracking until compelling new evidence is provided". The U-turn ends a potential dark chapter in the UK fossil fuel history. Fracking was once proclaimed by Boris Johnson as "glorious news for humanity" and urged the UK to "leave no stone unturned, or unfracked".

Back in the US, the Independent Petroleum Association of America claims that – "Fracking has been safely used in the United States since 1947. More than 1.7 million U.S. wells have been completed using the fracking process, producing more than seven billion barrels of oil". Fracking is not safe, not clean and another disaster for the environment. Toxic chemicals are used to create the 'frack fluid' which is pumped underground to squeeze out the oil. Frack fluid leaks into the surface water supply, in rivers and lakes. The vast quantities of water required for fracking are often in scarce supply in drill regions, requiring pumping and transport of water from elsewhere. Citing competition, many oil companies were allowed to keep the chemical composition of their frack fluid secret. A US investigation by Congress in 2011 found over 300 different chemicals were used which were also linked to cancer and other health risks. Frack fluid, for example, contains benzene, 2-butoxyethanol (used in antifreeze), bromide and has been known to kill vegetation and animals at drill sites. The climate impact of fracking can be detected in the atmosphere. Apart from the oil produced, fracking releases the climate killer methane. In 2019 a study by Robert Howarth, at Cornell University linked the massive increases in methane emission to fracking. Methane contains a signature that can be detected in the atmosphere.

In 2019 and 2020 a series of bans began to emerge on fracking. In 2019 the state of Oregon banned fracking. European countries such as Germany, France, Ireland and the Netherlands have enacted fracking bans. The days of Fracking are coming to a slow end. Another avenue for the oil industry is closing. Vice president Kamala Harris promised to ban fracking outright in her presidential bid. Almost 90% of new oil wells in the US, as of 2020, were Fracked wells. Increasing pressure and regulation will put tremendous pressure on the industry in the 2020s decade.

The next major threat is the industry itself. Built on a legacy of price-fixing to cushion demand highs and lows, organisations like OPEC are finding it increasingly difficult to stabilise revenue certainty. For over 60 years OPEC has represented major oil-producing countries in setting production levels to effectively create the world's oil price. Representing over 50% of the world's oil production and 90% of proven reserves, OPEC has had a dominant grip over Earth's major economies. The OPEC cartel caused the 1970s oil crisis with an embargo on supply following the Yom Kippur war, increasing the price of a barrel from around $12 in 1970 to over $30 by 1975. In the past ten years, oil has averaged around $50 per barrel. Cracks in the cartel's alliance are beginning as poorer members become increasingly aware that the oil game is potentially up sooner than planned. Rifts began to emerge between Saudi Arabia and the UAE over production levels in the months after the Covid crisis began. Whilst global oil demand is forecast to increase by around 1.4% per year (according to OPEC) the market share for energy and mobility is expected to drop. With more renewable sources of energy and the growing number of electric cars, oil is expected to drop below 30% by 2030, down from 80% in 2016. Without fracking and with its share of the market dwindling, another major threat to oil is discovery. The number of 'big finds' has been on a steady decline since the 1950s. The world's biggest find, the Ghawar Field in Saudi Arabia was the last. The number and scale of new finds continue to drop forcing oil companies to drill deeper, often in tough underwater locations, like the Deepwater Horizon's location in the Mexican Gulf.

However, in early 2021 a new Permian basin like discovery may have been uncovered in Namibia, Southwestern Africa, possibly up to 11 billion barrels of oil. Exxon has already acquired 7 million acres in the region whilst others like Royal Dutch Shell, Chevron and BP are all staking claims. The world's major oil companies have raced to the region like flies to the proverbial 'shit'. This is despite their public-facing claims to be transforming to become green energy providers. Old habits die hard in the hydrocarbon game, however, this is a temporary drug fix in oil's otherwise slow demise.

PARIS CLIMATE AGREEMENT

Within hours of his inauguration, President Joe Biden re-joined the US to the Paris Climate Accord reversing the previous administration's love of the fossil fuel industry. During the Trump years, the US took a step back from climate-friendly policies and renewable innovation. The Trump administration's pandering to Big Oil included reducing electric car incentives, relaxing environmental laws and lending his ear to Big Oil lobbyists, who were also often campaign contributors. These actions set the US back in the race to reduce carbon emissions. This period in American political history is now over.

Biden's green agenda with immediate executive orders overturned Trump's regressive approach. The Paris Climate agreement is an accelerator to the electric car transition and a further major threat to Big Oil. Under the United Nations Framework on Climate Change (UNFCCC) the Paris Climate Agreement is a legally binding treaty aimed at reducing CO_2 emissions and keeping the global warming increase to below 2 degrees Celsius of the pre-industrial age. The agreement works on five-year cycles of increasingly ambitious climate recovery steps. Although the agreement was signed in 2015, each country had until 2020 to present plans to reduce greenhouse gas emissions to reach agreements goals. From 2024 each country will have to produce transparent reports on actions taken and emissions reduction. These nationally determined contributions (NDCs) are up to each country to decide on which actions reduce emissions. Whilst each country has a different set of challenges, depending on energy mix and reliance on fossil fuels, almost all have begun to introduce a phase-out of internal combustion mobility.

For Big Oil, the Paris Agreement is bad news. A further attack on its business model by setting increasingly stricter emissions targets and pollution controls. Unlike the failed Kyoto accord, the Paris Agreement is the planet's best hope for a sustainable future as time is running out on global warming. The US withdrew from the agreement in 2017 citing grounds of competitive advantage. Ironically the US was already on track to meet its agreement targets, however political will evaporated following the Trumps election and the US regressed to the energy and mobility policies of the past. There are no penalties for breaking the Paris Agreement, it was built upon the mutual acceptance of the great global challenge facing humanity. The reasons for breaking with the agreement were dubious and unfounded. Trump made claims such as 'the agreement blocks the development of clean coal in the US' and the agreement would cost American's millions of jobs by 2025 and cut trillions off the US economy by putting the abundant US energy reserves under lock and key. Stating the US would be at a permanent disadvantage and ignoring the benefits of a cleantech shift and the economic growth from new energy technologies, electric cars and clean air health savings, he unilaterally withdrew from the agreement.

Big Oil lobbyists claimed the US was at a competitive disadvantage to other countries and argued that a level playing field needed to be created by leaving the agreement. The Big Oil majors spent $1bn on misleading climate-related branding and lobbying since the agreement was signed, according to Global InfluenceMap. President Trump's re-election campaign was flooded with Big Oil donations. Kelcy Warren, an oil pipeline billionaire, hosted a fundraiser, which Trump attended, raising $10m. Fossil fuel industry support in the White House fluctuates from administration to administration. Reagan removed the solar panels on the roof of the Whitehouse shortly after gaining office, while Bush Junior was a strident supporter of Big Oil.

With the Paris Agreement out of the way, oil companies approved over $50bn, in 2018/19, on new exploration projects, according to think tank, Carbon Tracker. In contrast to this, the oil companies continue to make ambitious carbon reduction statements in shiny annual sustainability statements. BP for example states that its new mission is to produce low carbon oil and gas in line with the Paris Agreement whilst continuing to make a massive investment in new fossil fuel projects. New chemical, oil plastics plants and oil infrastructure are planned in places like Louisiana's 'Cancer Alley' where a high concentration of oil refineries and petrochemical plants already exist. Cancer Alley is a predominantly African American and poor community – more easily bullied than affluent white suburbs and their NIMBY (not in my back yard) resident groups.

What was once a fringe subject, climate change, is going mainstream in the public's consciousness; it is slowly becoming unacceptable to doubt the science. Smoking in a public restaurant now is illegal and completely unacceptable today, yet driving heavy petrol or, worse diesel, polluting cars down a public street is still largely accepted or ignored. Big Oil and related industries are busy promoting three broad plans in their corporate boardrooms; 1. Sweating the existing oil assets they have for as long as possible, 2. Genuine divestment into cleantech and energy alternatives but also, unfortunately, 3. Delivering well-crafted public messaging to confuse the public and wash their hands of their ill-gotten gain. This is known as Greenwashing.

GREENWASHING

"In the interests of the environment please try and reuse your towels before placing them in the laundry." A common notice on a little card in your hotel bathroom you have no doubt seen before. It's a commendable request, use less water, use fewer chemicals, save our environment.

No, it's Greenwashing, a term coined by New York environmentalist, Jay Westervelt in 1986. The term has been used for decades when companies or institutions try and purport their activities or products are environmentally friendly by superficially promoting misinformation or advertising campaign. These token gestures may seem rather innocuous, however, Greenwashing on a grander, more insidious scale is increasingly taking place. Large corporations spend millions promoting their green credentials to mislead their customers. The Oil industry has been using this tactic for decades. A variant of Greenwashing is green scamming – a deliberate policy to use environmentally friendly-sounding wording on products or services. Ads for 'clean natural gas' are commonplace across the world for example, while one of the largest Oil organisations in the world, Saudi Aramco, has been awarded environmental awards (paid for) by the "The Green Organisation". I recently drove into a major highway service station forecourt in Ireland run by the AppleGreen organisation. A large advertising horde had been erected with a picture of company executives standing in front of some green plants. It read “Put Carbon into Neutral with AppleGreen”. Their website stated, “As part of our journey towards electric vehicles, and a low carbon future, we’re now offering you the choice of carbon-neutral driving with AppleGreen PowerPlus fuel”. A public social media row ensued, with the company being called out with direct accusations of Greenwashing.

In February 2021, whilst walking through Dublin International Airport I was confronted by lots of digital displays stating the airport had been awarded carbon neutral status. On closer inspection, it appears this has been achieved by switching to more efficient lighting and carbon offsetting by supporting African water projects. The airport also upgraded its ground fleet, effecting an overall 25% reduction in CO_2 emissions. A corporate tick in the box for the executive assigned with meeting ESG (environmental, social and corporate governance) goals.

The award was given to the airport by ‘a globally recognised environmental programme’. An image of six LEVs (Low Emission Vehicles) accompanies their claim. They are hybrid cars – more about the hybrid lie in a future blog. The accreditation was provided by the ‘Airport Carbon Accreditation’ – a programme run under ACI Europe – the European Airports trade body. Reducing carbon emissions by any clean measure is a step in the right direction and good news for global efforts. Greenwashing, however, by claiming to be carbon neutral through carbon offsetting, will not meet the goals of the Paris Climate Agreement and will not reduce global carbon emissions. Greenwashing through offset programmes is akin to burning tires in your back garden and planting trees in your front – then claiming you are neutral. Oil companies have been going all out on Greenwashing.

BP for example has committed just 3.2% of its budget for renewable investing while claiming to be on a mission to carbon neutrality by 2050. This does not mean any more oil and gas production in 30 years, rather it means investing and developing in clean technologies to offset their oil operations. Actions do not match the words of big oil with continuing investment in new exploration dwarfing investment in cleantech and renewables. Between 2020 and 2024 the oil industry is set to expend $1.4bn in new extraction projects according to the Global Gas and Oil Network. Exxon's 'algae-biofuel campaign' is a good example of greenwashing. The highly promoted project will equate to just 0.2% of the company's refining capacity. Exxon is allocating less than 1% of its revenue to developing clean energy. This false promotion was called out by a Massachusetts Attorney General, where she stated, "the company (Exxon) violates Massachusetts law through a deceptive 'greenwashing' marketing campaign that misleadingly presents Exxon as a leader in cutting-edge clean energy research and climate action". Analysis by Carbon Tracker, analysing nine larger oil companies stated in 2020 that "Most oil majors' net-zero strategies are greenwash, with US firms faring the worst"

Offsetting carbon emissions through new clean energy technology funding, planting trees and carbon projects are positive developments, relying on offsetting to try to neutralise still growing CO_2 emissions from the oil industry is not. The Oil industry knows this and knows that to exist in a future form they will have to make a massive transformative effort. Embracing the electric revolution is the first step in the right direction.

OILS GREAT TRANSITION

Big Oil's metamorphosis is underway from the slithering slug to the butterfly. Whilst they currently are only making relatively token clean tech investments it's the beginning of the end for their old ways. The oil companies are planning a $100bn divesture from their current assets. Shifting from oil to energy is their goal. Renewable energy technologies, battery technology, mobility, electric cars and autonomous vehicles are just some of the industries now grabbing the attention of big oil money. Energy transition and portfolio asset diversification have risen to the top of oil executive agendas. As the investment begins to flow to these new technologies the momentum will build and further accelerate EV and EV infrastructure evolution.

ExxonMobil, Royal Dutch Shell, Chevron and BP are actively pursuing new electrification and renewable strategies. BP, for example, acquired the UK's largest vehicle charging network provider, Chargemaster in 2018. This investment followed another BP entry into solar power with the investment in Lightsource. All the oil majors have now established venture divisions hunting green and renewable investment diversification opportunities. Making strategic decisions to move into these new industries could see big oil become big green energy tech. In a future world, they may survive and be the energy suppliers for electric generation through clean renewable technologies, such as solar, wind, wave and even nuclear fusion.

Survival is a powerful motivator and given the increasing pace of change, sitting still will not be an option. As the world's dependence on oil begins to wane, the oil tactics of greenwashing, misleading and false promotion of a dying product will be obsolete. There is a conflict between the strategic belief that oil will be around for a long time and the necessity to reduce carbon emissions today. Recognising the accelerating electrification of mobility, strategic execution has commenced for oil companies like Total and Shell. They have made commitments to cut carbon emissions by 50-60% by 2050. This required a huge sell-off in their most polluting assets, combined with a significant and immediate increase in alternative investments. Another key driver and motivation for this great transition will be investors. Shareholders and bankers are coming under increasing scrutiny to deliver on their green credentials. Greenwashing will only last so long, as the retail then institutional investors demand green transparency in their investing commitments.

Evidence of the transition is appearing almost daily with major announcements in the oil sector. BP announced it will develop 50 gigawatts of net renewable power generation by 2030, 20 times more than it produced in 2019. Low or no carbon producing technologies are becoming fashionable for the hydrocarbon gang to wear. The Biden administration is not as pliable to oil lobbying as the Trump one was, and oil companies know it. The transition needs to be genuine and not a cover story for the possible $4.9 trillion spend on new oil and gas production developments planned over the next ten years – according to Globalwitness. The oil industry could accelerate its transformation by diverting this enormous planned future exploration investment into a green transition.

The world has had a glimpse of what could be possible. The oil demand drop due to the 2020 pandemic cut greenhouse gas emissions by 6.4% - this is equivalent to two years of Japan's total CO_2 output, but unfortunately, it was only temporary. As the lockdowns were eased in the second half of 2020, the CO_2 emissions quickly rose again to pre-Covid levels. There is a growing concern and early evidence that the oil industry might try to mitigate their decline by unleashing a flood of cheap oil for the plastics sector. Despite the international plastics crisis with our oceans contaminated by microplastics – big oil could yet again be the environment's arch-villain. Every human is estimated to ingest nearly 2,000 microplastic particles per week from water, food and even the air.

In the past 50 years, the top 20 companies have contributed almost 500 billion tonnes of carbon dioxide to the atmosphere. Saudi Aramco at 60 billion tonnes leads the group with others like Chevron, Exxon, BP and Shell contributing 40,30 and 25 respectively. The oil industry may try to launch a dying last-ditch effort to keep the petrochemical world alive in the guise of lower carbon-producing fuels and cleaner refineries. With the world now on an unstoppable move towards zero-emissions mobility and clean energy they will have an uphill battle. The transition must happen and with Biden aiming to drive $1.7 trillion over the 2020's decade into renewable power and electric vehicles, through tax credits, research, carbon capture and investment, the potential for hope is real.

Concrete actions, like reducing emissions from cars, trucks and SUVs are in plan alongside President Biden's existing goals to create a carbon pollution-free power sector by 2035 and a net-zero emissions economy by no later than 2050. John Kerry's role as climate tsar is to save the planet. When asked about the necessity for the success of the new US climate plan Mr. Kerry said, "I'm not going to contemplate anything else,"… "because it would be a dramatic failure for mankind."

The Age of Oil is over and so too is the Internal Combustion Engine era.

CHAPTER 3

THE END OF THE ICE AGE

IN THE BEGINNING

Movement is the key to human progress. Humans largely stayed put for most of our existence over the past 50,000 years. There have been some mass migrations, usually due to climate changes, exhausted food resources or conflict, but in the main where we were born, we stayed and died. On a day-to-day basis, humans huddled in small settlements and rarely ventured a few miles from home to forage, hunt or plunder neighbours. Feet and fatigue limit the range of a human body to about 30 kilometres per day, however, that's a very expensive use of energy and can't be sustained for long. Staying in one place limited human development for thousands of years until we found ways to cover more distance with less energy. Discoveries, new foods, languages, metals and conflicts all came about when humans began to venture far from their origins. It wasn't until an unlikely relationship between human and beast resulted in a faster, less energy-sapping form of transport, facilitating long-range expeditions. We have our four-legged friend, the wolf, to thank for what would ultimately evolve into the modern car we know today.

Survival is the great inventor. The domestication of the wolf into sleigh-pulling machines was the start of the first journeys farther than human legs could walk in a day. Archaeologists have discovered that it was in the earliest alliance between man and beast, as far back as 15,000 years, that modern transport has its origins. In the far Siberian north settlements have been discovered that are over 9,000 years old. There, 500 kilometres into the Zhokhov Island, dog bone analysis proves they had been bred for sleigh pulling. Being able to cover long distances by dog sleigh meant the inhabitants could track down and follow fast-moving reindeer herds, on which they would have relied for food and clothing. Overland transport routes soon developed, and settlements began to interact. Bartering, toolmaking, and weaponry followed and then, so began human progress. Dog power ruled for thousands of years and so did ox power, then horsepower and soon kilowatt power.

The car took a long time to emerge. Before four legs could become four wheels the least applauded invention had to be made. The wooden axle. The axle is more important than the wheel. Without a perfectly circular axle to sit within the centre of a set of wheels, smooth, reliable transport is impossible. It took until the bronze age before sufficiently precise tools could be crafted to enable wood to be carved and smoothed to ensure a low friction gliding surface. The wheel (or rather the wheel-axle partnership) is relatively new, it only came on the scene around 4,000bc. The wheel has only featured in the past 6,000 of modern man's 50,000-year history. With the wheel came the real movement of people and goods. The sleigh was outdone by wheels on dirt tracks and then modern paved roads, pulled by horse for 5,000 years before the first steam engine was born.

Thomas Newcomen is credited with the invention of the first steam engine, built to pump water from coal mines in 1711, based on a 1606, Spanish design. What began to extract coal from mines more efficiently soon gave rise to the industrial revolution. Further refinement by Scotsman, James Watt instigated multitudes of engine designs, including the locomotive engine in 1802. Exploiting energy-dense coal, the steam train spread across the globe and for over 150 years was the king of transport. Steam began to be replaced from the 1950s by diesel and electric trains, the electric revolution began sooner for rail than it has for the road.

In 1886 Carl Benz successfully perfected earlier Belgian and Italian internal combustion designs to produce a workable two-stroke gasoline engine. At first, his machine was seen as a noisy, impracticable mode of transport. With a small range, limited fuel tank, lack of filling stations and smooth paved roads, the first automobile looked like a novelty for the few and an unlikely horse replacement for the many. The horse has a range of 80+ kilometres per day, early internal combustion cars came with range anxiety, not dissimilar to electric cars today. It took his wife, Berta, to prove the viability of the car by driving over 100kms in 12 hours to visit her parents. The story of her adventure soon spread, and the age of horse travel began its decline. What began as a slow trickle for the rich, with limited infrastructure, soon went exponential and within 20 years mass production of automobiles had begun with the Ford Model T.

THE ICE AGE

As with the steam age, the world quickly developed the necessary infrastructure to support mass automobile adoption. From the arrival of the affordable Model T in 1908, the US, for example, had a nationwide road network established in just 20 years. Gas stations, garages, tire services and highway rest stops sprung up almost overnight. Range anxiety was short-lived with a vast network of gas stations across the world. The doubters for this new age gradually faded away and the internal combustion engine became king in less than 25 years. The car soon became the ultimate freedom machine. Cars have since terraformed our world with over 32 billion kilometres of road now crisscrossing the earth today, but the ICE age was almost the electric car age. Electric cars came first.

In as early as 1832 the electric carriage was born by Scot Robert Anderson. These crude first attempts never gained traction and others flirted with electric motor driven wagons for a few decades after. Rechargeable batteries came on the scene in 1859 with the first lead-acid batteries. By 1900 one-third of all cars in America were electric. Electric cars were relatively low speed, high cost and low range in comparison to the ICE cars that followed. Thomas Parker built the first production electric car in 1884 in England after becoming concerned about the effects of pollution having electrified underground London railways. 110 years before Tesla Cars was born, a fleet of electric taxis roamed the streets of London. In New York too, electric cars were introduced in 1897 with 62 Hansom cabs. Despite a limited range and lack of national electricity networks, which largely confined electric cars to cities, over 30,000 electric cars were registered in the United States, outselling steam and gasoline-powered vehicles by 1910. A form of battery swap subscription service was pioneered in the US to try to overcome the limitations with range and lack of charging infrastructure. Thomas Edison developed an electric car in 1912 with a range of 190 miles. The nickel-iron battery that powered it, however, was expensive to produce compared to a gasoline engine, making the car cost almost three times more than the Ford Model T, so it failed.

The mass discovery of oil reserves made the availability of cheap gasoline widespread. This, coupled with the newly built road networks and affordable gas cars, sealed the fate of the fledging early electric car industry. Henry Ford worked for Edison, but then his own mass-produced, simple to maintain car sold 60 million units and changed the world as we know it. By the 1920s the electric car was dead. Apart from small delivery vans, golf carts and milk floats the electric car faded into obscurity for the next 50 years. It made a brief comeback in the 1960s with early hybrid concepts and prototypes that never really gained public interest or made it to mass production. The electric Lunar Roving Vehicle or moon buggy, three in total, covering over 100km on the surface of the Moon during the Apollo missions in the early 1970s, the first cars to be sent into space, (until Elon Musk launched his early edition Tesla Roadster on board one of his Space X test flights in 2018).

With the low cost of gas, the incredible multi-lane highways and huge competition from a multitude of car manufacturers, the ICE age exploded. Car ownership grew exponentially to a global peak of over 1.4 billion vehicles today, less than 1% of them electric. Our world has been fundamentally shaped by the ubiquitous car. Almost every aspect of life is centred around moving. Our cities are shaped by cars; streets, pedestrian systems, car parks, homes, workplaces and laws have been dictated by the car. The ICE age has spawned everything from motorsports, accessible holiday escapes to oil and service station empires. There are over 110,000 gas stations in the US alone. The top ten largest oil companies own and operate over 200,000 gas stations worldwide. Building, maintaining and servicing ICE cars has created whole industries and complex supply chains across the globe. There are almost four billion tires produced each year globally. The world's three largest economies, the US, EU, and China scrap almost 25 million cars per year. Modern life today simply couldn't function without cars.

However, whether you build, sell, own, rent, or share a car, the ICE world as we know it is rapidly coming to an end. It is not just the replacement of the drive train from gas to electricity that will be different; whole industries and businesses are about to disappear. New opportunities are being created and new challenges are emerging. Sometimes the wrong technology wins, and the electric car was largely forgotten for a century, until now. With the environmental necessity and the electric resurrection, the very essence of our interaction and use of the car is about to be radically altered and nobody is immune.

INEFFICIENT INGENUITY

If aliens visited Earth today, stopping by on the moon first and spotting a 50-year-old abandoned, electric NASA rover, they would probably conclude humans had taken the 'normal' route in development and evolving self-propelled mobility by harnessing electricity for motion.

When they got to Earth and witnessed the millions of noisy, polluting and fossil fuel burning cars they would likely baulk in disbelief and wonderment. Aliens would marvel at this complex mechanical contraption. The modern internal combustion engine is an incredibly complex and yet completely inefficient machine. The propulsion system explodes hydrocarbon fuel and uses the energy produced to cool the machine to stop it from melting and then, uses the remaining energy for motion via a convoluted and intricate set of reciprocating parts and gears.

The ICE car is on par with a refined mechanical Swiss luxury watch with hundreds of tiny components each interacting with each other in a delicate balance. The ICE car does not make sense, it is an illogical option for humans to choose as their primary means of transport. The energy produced when gasoline is exploded is over 80% lost with internal combustion engines, with electric cars this loss is just 10% - 15% The bicycle is more efficient at converting energy into motion than an ICE car. ICE cars have around 2,000 moving parts compared to electric cars with just 20. ICE cars are complex, require multiple parts to be maintained and wear out faster. Although the current battery density of today's electric cars delivers less energy per kilogram of weight than gasoline, the use of that energy is hugely more efficient than combustion. On average, electric cars are five times more efficient than gasoline equivalents. ICE cars simply lose huge amounts of burned gasoline energy before it drives the wheels. Cooling the internal explosion mechanism (thermal loss) and exhaust heat loss loses 60%-70% of the energy immediately. Another 10% is lost from the various systems draining power such as ignition systems, water and fuel pumps. Idling loses another 3% while the drive train (gears, transmission) loses another 5-6%. What's left is the power to the wheels. However, the waste doesn't stop there. Once the car is moving it must be stopped, but conventional friction braking is counterintuitive, converting the hard-won kinetic energy into heating the brakes and tire wear to stop the car. Burning energy to make it move then wasting more energy to make it stop.

Karl Benz's internal combustion engine has come a long way, there have been many innovations and refinements over the past 120 years. The basic principle remains the same however, air & fuel are mixed, exploded, a piston moves then gears, cranks and drive shafts move the wheels. Some of the biggest developments include fuel injection systems. The 'i' that appeared on many 1980s hot hatch models stands for fuel injection. This innovation dates to World War Two aircraft. Fuel injection eliminated the carburettor and pesky choke that older readers will remember. Fuel injection is more efficient, less troublesome and simpler to maintain. In the 1960s came the 'turbos'. Again, dating back to World War Two airplanes, the turbocharger is a snail-shaped turbine that compresses more air into the cylinders producing more power at the ready.

Variations on internal combustion followed like rotary engines with their oval-shaped chamber and triangular rotating 'eccentric shaft'. Adopted and evangelised principally by Mazda for years, rotary engines have fewer moving parts, however, still work on the same principle of a mix, compress, bang, and blow. Apart from the rotary engine, which wasn't widely adopted by car manufacturers, the remaining changes from Mr. Benz's design are the configuration of the pistons. Starting from the Model T's 'straight 4' layout, engineers developed a variety of configurations over the decades, up to V12s and even V16s, but otherwise little else has changed mechanically. A host of electronic management systems and fault detection sensors have been added over the years making the combustion more efficient and relatively cleaner, however, the other limits of ICE innovation have been reached, innovation has plateaued. With an electric car, 80% + of the power from the battery is delivered to the wheels. Tesla still holds an efficiency advantage over all other electric cars on the market today – more about this later- however, even the early electric cars of one hundred years ago still produced more movement efficiency than today's gasoline cars. 80% of the oil humans brought to the surface, refined and burned in ICE cars for over 120 years was wasted. 80% of the climate-destroying CO^2 in the atmosphere from ICE cars, was for nothing. The motor and oil industry engineers have known this since the beginning, but vested interests, the relative ease of oil extraction and suppressing the electric car ensured that gasoline cars thrived. In June 2021 the EU confirmed this with a $1 billion fine on German automakers for colluding to hold back technology developments that could have reduced emissions.

Enormous human effort and resources went into researching and developing the energy-wasting combustion-engine car and only a fraction of that effort into the efficient electric alternative. Overall, the

ICE car is a pathetic use of energy, and we destroyed our climate by prolonging it. This would make zero sense to the aliens, and they would probably conclude that humanity was an irrational, underdeveloped species and reschedule a date to check back in on us in 1,000 years, to see if we made it.

THE PROBLEM WITH CARS

Freedom created by the car brought prosperity, people and the world together. The ICE car has been overwhelmingly successful – one of the top 10 selling products ever produced. However, success has come with a price. I have covered climate damage already, but the car has caused a multitude of other impacts. Cars invaded and utterly transformed city centres. What were once quiet streets, save for the sound of horse hoofs, because noisy and crowded. City streets, with fast-moving automobiles, became dangerous. Cities are unrecognisable since the dawn of the car. Traffic lights, parking meters, road signs, painted road markings clutter the thoroughfares. Traffic, serious congestion and human frustration are a feature in every major city on Earth.

We are used to noise pollution and have tuned it out when walking on a busy city street sidewalk. It is only when you try to speak with someone, try to listen on the phone or on headphones that you realise just how incredibly noisy ICE cars are. I never really paid too much attention to this until I converted to an electric car. When an ICE car accelerates past me now, knowing the noise is the by-product of 80% wasted energy together with a dose of air poisoning and carbon emissions, I get a little angry. With way more impressive acceleration, an electric car can do the same and more with little noise and zero emissions. After congestion, noise and pollution, the next major issue is parking. Every driver has had the experience of trying to find a parking place when late for an appointment. Overpriced parking lots, parking fines, wheel clamps, tow trucks and parking wardens are the antithesis of freedom. Over the decades the car has developed a bad name. Anti-car lobby groups, city restrictions and even congestion charges designed to push cars out of cities are now commonplace.

When I was 17 years old, I had the single dream to have my own car. Living in the Dublin suburbs I would imagine my future with total freedom to go wherever and whenever I liked. My friends all thought the same way; getting a car was everything. I had a poster on my bedroom wall of a BMW 3 Series, E30 model (the boxy 1980s shape) that read 'the ultimate driving machine'. With summer and after school jobs I saved up and finally realised my dream, buying a used 3 Series when I was 19. However, owning a freedom machine, for young city-dwellers, today is no longer an attractive option. The total cost of ownership with parking, fuel, insurance and depreciation doesn't appeal to new generations. 64% of millennials surveyed across multiple European cities said they have no intention of ever owning a car. This is different for suburbanites with young families, but the trend is overall the same; people are turning away from car ownership. Ridesharing, car sharing, peer-to-peer (Airbnb-style own car-sharing) and new leasing models now offer the driver alternatives to buying a car. The electric car revolution is happening in parallel to these new ownership models and many ways enabling a new era of mobility. Electric cars are being developed with future shared ownership in mind.

Electric cars alone don't solve all the other problems associated with the car but they are acting as a catalyst for change in how we move. The electric car revolution is not just about swapping gasoline for a battery, there are seismic shifts taking place in parallel. Ownership behaviours, changing cities and autonomous cars will mean not just an electric future but an entirely different transport experience in the decades ahead.

THE BEGINNING OF THE END

The rise of Tesla resurrected the electric car. From producing a novelty electric roadster in 2008, Tesla has become the world's most valuable car brand in just 12 years. What started as a ridiculed start-up, declared 'destined to fail' by the experts, and run by people who knew nothing about building cars has now shaken the automotive world to its core.

The success of Elon Musk's Tesla at first didn't worry the ICE car makers, they foresaw a near-endless future for gas cars. That has all changed. The electric car has created a revolution that is increasingly accelerating the end of the internal combustion engine. With the clock running rapidly down to climate disaster and governments scurrying to action, the transition to EVs has been catapulted by ICE bans announced by countries and cities across the globe. Like the smoking bans introduced first in Ireland, it is quickly gathering pace and spreading across the world. The writing is on the wall for internal combustion, ICE cars are in the departure lounge. Most manufactures have begun the transition to electric in a late and desperate race to avoid becoming the next Blackberry or Nokia to Tesla's 'iPhone'. OEMs were sleeping at the wheel and Tesla was virtually unchallenged until now. Until very recently, the existing car manufacturer's first EV attempts at competition were mutton dressed as lambs. The only way they will survive the next decade is to create separate stand-alone EV units, new brands and wind down the ultimately bankrupt legacy ICE businesses. Research and development have all but stopped for ICE. In September 2019, Daimler and VW declared an end to developing the next generation of internal combustion engines. The change will be massive for the industries that supply and maintain ICE vehicle production lines. There are clear indicators that the age of ICE is near death and that the revolution is happening faster than most people are aware.

Nearly all the major OEMs are investing billions of dollars in electric vehicles. Ford, for example, announced an $11 billion plan for up to 40 EV models. The OEMs are waking up, and so too is the battery industry. The cost of battery production has been consistently dropping and with new production facilities, like Tesla 'gigafactories', worldwide supply chains are being established. A huge variety of new charging infrastructure projects are underway and planned to eliminate 'range anxiety' and hasten the consumer's transition. The cost of electric charging versus gasoline is lower and the total cost of ownership of an EV is now (as of 2022) on par and quickly becoming lower than owning an ICE car. Research and development on new battery technologies like solid-state, faster charging times and longer ranges will quickly overtake a century of internal combustion innovation. Electric cars are also more reliable. Fewer moving parts need replacement and servicing. Batteries last a long time, cover a lot of miles and do not degrade anywhere as quickly as some of the misinformation out there suggests.

Electric cars now also have a better residual value over their ICE counterparts. With the coming ICE bans, a deluge of new electric models coming on the market and better ownership economics. consumers will shift en masse by 2025. This will further accelerate the depreciation of ICE cars with a likely value collapse and a possible glut of unsellable used ICE metal after 2025. Finally, the driving experience is way better in an electric car according to surveys – 91% of EV drivers in the UK for example say they would never go back. That number is 82% for US drivers, according to a January 2021 poll by JD Power. In 2020, Europeans bought more EVs than the previously loved diesel car according to JATO Dynamics. Existing car plants, supply chains and workers now face an uncertain future. In early 2021, General Motors announced a strategic decision to phase out ICE car production and only make electric by 2035. Ending a century of ICE production, unfortunately, casts a shadow on the future of GM's 160,000 workers. ICE manufacturing is mostly about the batteries; the need for components like clutches, cooling systems, gearboxes and pumps are obsolete and as the EV revolution gathers pace these jobs could become obsolete quicker than workers may expect.

BANISHING THE BEAST

On September 23, 2020, Governor Gavin Newsome announced a Californian ban on all Internal Combustion Engine Vehicle sales by 2035. Californians have been dealing with increasingly deadlier wildfires and record temperatures year on year. Newsome's move was at odds with then-President Trump's fossil fuel/big oil-supporting policies. Banning ICE cars is part of California's state-wide goal to be carbon neutral by 2045. That commitment by the world's 5th largest economy, in a country of gas-guzzling SUVs and pickup trucks, was a huge boost for the electric revolution. New Jersey soon followed, then Massachusetts in January 2021. These US pioneering states follow several European country and city announcements. Norway, which topped over 60% of new cars sales being electric in 2020, led with a 2025 ban on new ICE cars. The Netherlands, UK, Ireland, Germany and Austria have all announced new ICE sale bans to come into effect from 2025-2030. In November 2020 the UK government brought its ICE ban date forward to 2030. Some countries, like France and Spain, remain outliers with 2040 target dates. Globally, major economies such as Japan and Korea have also announced bans planned for 2035.

As with smoking, it is virtually certain that every country in the world will eventually place a ban on ICE cars. While these bans are for the sale of new combustion cars, it doesn't mean they will be 'illegal'. In practicality petrol and diesel cars will be still on the roads well into the 2040s, however, will gradually become more and more scarce on the roads. However, several cities are introducing outright bans on ICE cars. Amsterdam announced that no internal combustion cars will be allowed in the city after 2030, Brussels announced 2035 and Hong Kong wants to completely phase out all gasoline cars by 2040. The pace of the electric revolution has been underestimated by governments and city officials worldwide. Almost all predictive models assumed a linear transition. However, a survey in mid-2021 by Transport and Environment, in over 15 European capitals, revealed that over 60% of residents want zero-emissions vehicles only by 2030. This is the same pattern whenever new disruptive technology is introduced. Just as with the steam era, the internal combustion age, electrification, mobile cell phones and many more examples, once a new technology is introduced that is better than the incumbent or beneficially transformative for humans, its adoption accelerates exponentially.

In September 2017, I attended a conference in Ireland on the challenges electric cars will bring and the government's plan for rollout. I was scheduled to speak to the gathering of car park owners, city planners, energy managers and independent businesspeople. Just before my turn to speak, a government lady, responsible for charging infrastructure planning, gave a presentation on the expected take-up of electric cars. Her chart showed a linear transition out to 2040 where the government expected just over 20% of all cars to be electric. There was no reaction from the audience. When it was my turn to present,

I respectfully contradicted her projections. I explained the indictors and trends pointed to electric cars going exponential. I suggested that by 2040 'all' cars would be electric and by 2030 no one would be purchasing an ICE car. The audience reacted with disbelief. In the bar after the event later than night, several delegates made a few friendly jokes about my statement. That was late 2017. Within two years major city administrators and governments had announced ICE bans planned before 2040 and some of the biggest car manufacturers declared their future was electric. If I was to make that same statement, to the same audience, in 2022, no one would disagree, such is the accelerating pace of the transition.

The EV boom is accelerating. 20th-century business models will fail in the 21st century. The ICE car was designed for success in the 20th century. There is no longer any rationale for continuing with gas cars. As the public becomes more aware of the electric choice available, and as governments enforce gas car bans, the electric age will begin. ICE cars, like steam before, will be museum pieces and novelty experiences.

Climate change and the rise of Tesla sparked the beginning of the end for Big Oil and the internal combustion engine.

CHAPTER 4
THE RISE OF TESLA

2003-2008

During November 2021, the market capitalisation of Tesla reached over $1.1 TRILLION.

That makes it the 6th largest company in the world and more than the next TEN biggest global car brands combined! Yet Tesla produces only a fraction of the number of cars compared to these global brands. Toyota and Volkswagen alone delivered almost 20 million cars annually compared to Tesla's 932 thousand units in 2021. Tesla released its first production car in 2008, the Roadster, and completed a successful IPO becoming stock market listed in 2010 with a market capitalisation of $1.7bn. In just 11 years the value of the company has increased by 700 times. In contrast, Ford, for example, was valued at $30 billion in 2010 and $81 billion in November 2021. Ford Motor Company, founded in 1903, is now worth just 7% of the value of Tesla, after 117 years in business compared to Tesla's 13 years. Tesla is not an internet stock, whose valuations can skyrocket overnight on the potential of exponential online user growth. Tesla makes a physical product, is constrained by the limitations of production output and to date has put just under two million cars on the road.

So how did this once ridiculed upstart manage to overtake a century of automotive progress in less than a decade and create the richest man in the world, Elon Musk? The answer is simple and complex at the same time. Vision, belief, singular focus, timing and growing awareness of climate change combined to create the iconic brand. Tesla not only rose from zero to hero in a decade, however, it has also shaken a century-old industry and sparked the most radical transformation of mobility since the Ford Model T. Tesla is responsible for resurrecting the electric car and bringing ideas previously considered science fiction ideas to reality; from Superchargers to self-driving cars, Tesla has created the future and it's just the beginning.

In July 2008 Elon Musk took delivery of the first Tesla Roadster, a modified Lotus Elise stripped down and packed with lithium-ion batteries. Capable of over 200 miles range and supercar-like acceleration of 0 – 60 mph in just 3.7 seconds – it was the first highway-legal electric lithium-ion production car ever. Costing over one hundred thousand dollars, almost 2,500 were sold over 4 years, mostly as expensive novelties to the rich and famous. Elon Musk's Roadster is now in outer space, where it is projected to orbit the sun for at least the next 20 million years.

Although most major manufacturers have declared their electric future and produced some respectable Tesla competitors, Tesla is simply still way ahead in efficiency, Supercharging networks, battery technology, software and data. It must be said that Tesla's meteoric valuation rise is in part driven by mob mentality and a degree of 'follow the crowd' speculative investing. Yes, there is a fair bit of irrational exuberance but… there's another reason. Perhaps the world is beginning to realise, just maybe, that Tesla was on to something after all. Just maybe, the electric car is the future, just maybe, the model of internal combustion is over and soon. Perhaps, finally, the way we have burned oil to move for the past one hundred years is about to suddenly come to an end. Perhaps the valuation is justified, and the other car manufacturers are struggling for survival as the electric revolution turns exponential.

Elon Musk did not start Tesla Motors. Founded in 2003 by Martin Eberhard and Marc Tarpenning, the South African immigrant became chairman in early 2004 following his $6.5 million investment in the company. After a legal settlement in 2009, there are five individuals allowed to declare themselves as Tesla's founders: Musk, Tarpenning, Eberhard, Ian Wright and J. B. Straubel. Only Musk remains at the company today. Tarpenning and Eberhard had researched the potential of electric cars and other propulsion methods for years before Tesla, having together developed consumer electronics in the past. They also founded the 'Mars Society', two self-admitted 'space-geeks' promoting a future human civilisation on Mars. It was at a Mars Society gathering that the pair first encountered Elon Musk.

Before Tesla and SpaceX, Musk started X.com in 1999, a 'fin tech' at that time in the payments space. Following a merger with what eventually became PayPal, his exit from the business netted him $165 million when PayPal was acquired by eBay in 2002. The pair contacted Musk and pitched the idea to him. Musk got the electric car vision immediately. The key was the lithium-ion battery. Old electric car concepts, with lead-acid or nickel batteries, would not allow for rapid charging and continuous re-use. At the time lithium-Ion batteries had a poor reputation over safety with their potential to overheat and explode due to what is known as thermal runaway. With engineering backgrounds, Eberhard and Tarpenning conducted thermal runaway experiments, inventing a battery cooling system and ultimately created a stable lithium-ion battery pack configuration. The pack was created by 'wire-bonding' multiple cylindrical battery cells together. This is effectively many single battery cells, not much bigger than a typical 'AA', wired together in an efficient configuration. Before their invention, no one had ever tried to put thousands of battery cells together to form a single battery pack. This method has become the standard for all subsequent battery pack designs. However, it's not so much the individual batteries, but the battery management system controlling temperature, discharge and recharge that makes it viable. This 'pack' is what gave rise to the viability of the new electric car with enough capacity for a credible range.

Musk saw the potential and the rest is history, albeit history still very much in the making.

ROADSTER

Although Musk invested the majority of Tesla Motors Series A funding round in 2004 and became chairman, he never set out to build an electric car, however, his conversion happened almost immediately, becoming very actively involved in the design and production of the Roadster. The Roadster prototype was unveiled in 2006, attracting a large curious crowd, held in a tent. Arnold Schwarzenegger, then governor of California, made a surprise appearance for the unveiling giving the team a massive publicity boost. Pre-order sales were via an 'owners club' – with no real marketing structure or plan in place. The company didn't have precise delivery dates and relied on an evangelistic follower base of 'believers' in the new electric project. The delivery dates moved constantly as the team struggled with supplier delays for many components, from seat belt suppliers to door handles and airbag manufacturers etc. This was mainly due to being an unknown and given their small production numbers many parts makers simply ignored the start-up.

In October 2008 Musk became CEO a short period after Eberhard became the first of the five 'founders' to exit Tesla Motors. Eberhard's exit from Tesla was anything but consensual or smooth. Although he was one of the visionaries and original designers for the Roadster, he and Musk had an acrimonious divorce with numerous lawsuits and an ultimate settlement. The Roadster came close to death on several occasions with transmission issues, reliability problems and production overruns. After five years in the business, Tarpenning followed his original co-founder and exited Tesla in 2008. Their original business plan was to follow the Roadster with a large Sedan or SUV to be followed by lower offerings. Their original plan was largely followed in the years later, even though they were not responsible for executing it. The Roadster changed the perception of electric cars, that part of their vision was achieved, however it was Elon Musk that was prepared to take the financial and technology risks to drive Tesla to where it is today.

Jay Leno, the TV talk show presenter and car collector interviewed Musk and test drove the Roadster back in 2008. When Leno questioned Musk on the use of a normal battery from fully charging and fully draining, referencing his golf cart battery-powered, original 1909 Baker Electric Car, Musk's answers annihilated battery preconceptions. Introducing battery management systems, he explained how it didn't matter what level of charge the car received and that was the game-changer statement. The Roadster had 6,831 individual batteries, a similar number to later Tesla models, all working together in a safe, temperature-controlled environment. The big breakthrough in using multiple cells instead of a single giant battery was the ability to 'ignore' a failed cell. This redundancy mechanism means that as cells fail the overall impact on battery output is minuscule. It would need over 700 cells to fail before 10% of the battery capacity is no longer useable.

The Roadster was a complete rebuild around the battery pack and motor. Lightweight panels were constructed, and the original Elise's was lengthened. Less than 10% of the original Lotus remained, the Roadster was its own car, revolutionary, ground-breaking and like nothing before. As with almost all new technologies, scale brings cost reductions and mass production brings scale, however, the Roadster was not built for the masses. Costing over $100,000 it was very much in the luxury market. The Roadster did destroy the pre-held perception of electric cars, it was cool, extremely fast, practical to charge and use every day. The addition of an induction motor, which allows the wheels to charge the batteries as it slows the car, known as regenerative braking, was another innovation meaning less brake wear and less waste of energy. The 80% energy lost by ICE cars had just been blown away. The Roadster delivered instant power to the wheels with incredible, no delay acceleration.

In contrast, ICE cars will never match the instant acceleration of an electric car. Despite over 100 years of development and refinement, the process of burning gas and converting piston movement into crankshaft rotation is just too slow compared to the speed of light electricity. Leno was taken aback by the performance, silence and range, concluding that if this was the future, he was not worried. In 2019 Tesla announced the revival of the Roadster name - a slick new sports model with a claimed outrageous 1.9 seconds 0 to 60 mph acceleration (considerably quicker than the $3M Bugatti Chiron hyper car) and a 250mph top speed. It was the original Roadster that began Tesla's rise to fame, its dominance in EV and a near-cult following. Not everyone was convinced of the potential of the EV upstart, however. The Silicon Valley gossip magazine, Valley wag, got it wrong when in 2007 they declared the Tesla Roadster as its Number 1 fail for a technology company that year...

BIG SEXY VISION

Musk likes grand visions, setting near unobtainable goals, he dreams big. Founding Space X, Solar City and part founding Tesla are all part of a 'master plan' in his head to drive humans to become a multi-planetary species. He may be right. With our poisoned air and slowly decaying climate, a plan B home would be a good option to have. His stated mission is to create a human settlement on Mars with a population of one million.

His Space X rocket company has already transformed the cost to payload ratio – making it the most competitive satellite launch company on earth. His 'Starship' prototype test rockets are the forerunners for what will ultimately take humans to Mars in the second half of the 2020s decade – according to Musk. Critics note the costs and the technical impossibility of his dreams, but the world has been largely shaped by dreamers, even if they only achieve half of their grand visions.

I had expected to vacation on the moon by now when I was a kid. I have always been fascinated by space. In 1978 when my mother brought me to see the first Star War movie, A New Hope, one of my few enduring childhood memories - I can recall queuing outside Dublin's Adelphi Cinema – to see this much-hyped space epic. I was blown away entirely and immediately with the opening scene, the giant Star Destroyer appearing from the top of the screen in pursuit of the Rebel ambassador's spaceship. So began a life-long obsession with everything space. The highlight of my obsession to date was being lucky enough to witness one of the last space shuttle mission launches in 2010, STS 132, Atlantis. Standing in the sweltering Florida heat, I was again blown away when the rocket boosters ignited and then the wall of sound and vibration hit me just after the shuttle began to rise. Fifty years earlier men walked on the moon for the first time. The rate of space progress from Sputnik, in 1957, to the first space shuttle launch, STS-1 in 1981 with Columbia was a staggering testament to what humanity could achieve. In just 25 years humans went from a basketball-sized metal sphere to walking on the moon and a reusable spacecraft, carrying a crew of seven humans into space. I was massively disappointed in the years since. The end of the cold war, focus on budgets and increased risk aversion ended the Space Race. Mars rovers like the successful Perseverance mission in February 2021, the International Space Station, the Hubble space telescope and the discovery of exoplanets have been astonishing achievements, but we have comparatively stood still for the past quarter of a century. It is now unlikely I will get to holiday on the moon or Mars, however, my children may, and like the rebirth of the electric car, the rebirth of the space race is partly down to Musk's daring ambitions.

Despite the relative success of the Roadster, Tesla was still heavily loss-making, but Musk set out the SEXY grand plan. His plan was simple to explain and, in his words, stated as:

1. Create a low volume car, which would necessarily be expensive
2. Use that money to develop a medium volume car at a lower price
3. Use that money to create an affordable, high-volume car

   And...

4. Provide solar power.

The MODEL S, 3 (because E is used by Mercedes), X and Y formed the grand plan – S3XY. The Tesla Truck, Cybertruck and a new 25k Tesla now also form part of the range bringing seven electric models to the market. Musk is not afraid to take risks - sinking practically all his windfall from PayPal into Tesla. The Model S – a sleek, luxury saloon designed by Franz von Holzhausen and inspired by the modern Aston Martin brand, is the flagship of Tesla. This ground-breaking vehicle was a step-change from the Roadster. An eye-catching car with eye-watering performance, 100% electric with a minimalist sparse interior. The centrepiece is a massive 17-inch touch screen (this was designed before Apple's iPad had even hit the market) challenging the very core of how a car's cockpit had been designed for a century.

On first sitting in the Tesla Model S, you are struck by the lack of buttons and dials, even handrails and coat hooks are absent. Like the near-empty 1900s electric vehicles, such as Leno's Baker, the S has just what it needs and nothing else. At least that is how it looks before you put your foot on the brake (the brake press 'wakes' the always-on car up). The giant touch screen comes to life and the smaller centre dash also illuminates. After the battery technology, the user interface, via the touch screen, is the heart and unique differentiator over all other manufacturers, even as of 2022. The centre console controls everything, from lights, suspension, heating, windows to steering wheel responsiveness, charging and regenerative braking. Steve Job's life philosophy was to develop consumer electronic products that were beautiful and simple to use but not simplistic. This balance is echoed in Tesla. The user experience, or UX, is extremely intuitive – after just a few minutes of tapping around you get it. No other explanation is necessary unless you never used a touchscreen or smartphone before. What's even better, and just like your smartphone, it's constantly upgraded with new features and improvements via the cars always connected internet. No need to visit the repair shop for technical issues, problems could be solved via over-the-air updates and software bug fixes. All of this in a car from 2012; in 2021 most EV competitors are still struggling to replicate this level of ease of use and functionality. Contaminated by their legacies, the existing OEMs are now trying to develop a Tesla-like experience in their new cars, however, are still polluting the experience with the habits from the ICE age; more about this later.

Save for a front nose job, the design of the Model S hasn't changed for almost a decade. Its sleek design has stood the test of time. I have driven a Model S for over four years all over Europe and Ireland. It is still a head-turner, especially with kids and teenagers. Every time I pass a group of teens they always look and often point – this does not happen with other manufacturers, even with their new electric cars. Tesla has the sexy brand appeal; these future car buyers of tomorrow are already hooked. What has changed is a host of features and functionalities thanks to the regular software updates. Battery preheating to maximise supercharging efficiency, a new entertainment module, complete with retro 1980's arcade games and Netflix (for use when stopped). Navigation is simple and intuitive, using a Google Maps interface. You can search for a business address for example without having to search for streets and postcodes.

When I accidentally destroyed my Model S wing mirror reversing into a garage I brought the car back to Tesla for repair. They were out of Tesla courtesy cars so gave me a rented Mercedes C Class diesel for a few days. When you get used to simple it's hard to go back to the complex. The navigation is near unusable compared to the simple Google Maps interface in Tesla. This brand-new Merc drove me nuts trying to turn the physical dial interface to search for streets, like trying to crack a safe from a crime movie compared to the tap-tap-go on the Tesla big screen. Following a suggestion by a customer to tone down the alert noises when his kids were sleeping in the back seats, – 'Joe Mode', named after the customer, was introduced and is now a feature. Fart mode, cowbell mode and continuously evolving autopilot features have all been added over the air. The Tesla Model S and Model 3 scored a 5-star safety rating by Euro NCAP. There is mostly space under the hood creating a large crumple zone. There is also the low centre of gravity with the evenly distributed floor weight, thanks to the heavy battery pack in the floor. The battery pack is also strong giving side-impact rigidity. On top of the physical safety, there is then the autopilot, activated or not it is constantly watching road conditions and alerting drivers to dangers from its multiple cameras. (more about autopilot and autonomous vehicles later).

The car has an adjustable air suspension that can automatically raise and lower the car based on exact locations from previous memory experience. The drive experience is like no other. A different feel from an ICE car. The acceleration is breath-taking, instant power delivered to all four wheels and without the noisy fuss of gas pistons screaming to accelerate. We have been conditioned to think this noise of a powerful ICE car accelerating is cool, I thought the same. I now think this noise is dumb. The Model S is not silent, at low speeds noise is barely audible but when you put your foot down there is a spaceship-like increasing power plant hum. It's the strong silent type. The X is a giant futuristic SUV. With electric, DeLorean style, gull-wing doors and a huge interior, the X is built on the same 'skateboard' as the Model S. The skateboard comprises the entire floor plate of the car, containing thousands of battery cells, drive motor and induction motor. This skateboard configuration is now the default for almost every electric car. The Model S and every Tesla since is hyper-efficient, lightweight and aerodynamic to offset the weight of the battery pack and increase range.

Both the Model S and Model X are luxury class vehicles. They were not intended to be the cash cow, profit makers. That title was reserved for the Model 3, the much anticipated 'affordable' electric car. Musk's original vision was to create a $25k car for the masses with the 3. That dream was not realised, in part due to the still relatively high cost of battery production and the lack of a matured production process with large economies of scale. Despite trimming out the extras in the Model 3, like electric door handles, internet browser and air suspension the car cost $35k for a base model as of 2021 (with government purchase subsidies) in the US. In Europe the standard range plus costs as much as €45-€50k in most countries with subsidies, keeping the Model 3 in the luxury segment. The 3's interior is even more spartan than the S with a single centre horizontal touchscreen and no traditional dashboard. It has been a huge success, earning Tesla the title as the fastest-growing brand with half a million pre-orders before going on sale. The Model Y completed the SEXY vision, essentially a mini-SUV variant on the 3. Almost one million Model 3 and Model Ys were sold in the first 3 years bringing Tesla to eventual quarterly profitability and catapulting the brand to global stardom. Getting to this position, from an obscure Californian electric car start-up to one of the most recognised brands in the world didn't come smoothly. Tesla and Musk have had a baptism of fire scaling up production with many serious setbacks and near bankruptcies along the way.

FREEMONT

To realise the S3XY dream Tesla needed to seriously upgrade from its small Roadster factory in Menlo Park, California. Belief in the new car company and cash were scarce when Musk and co went looking for a new car plant – they had to be creative. When Daimler acquired a 10% stake in Tesla in 2009 and Tesla management managed to convince the US Department of Energy to provide a $465 million loan to the company, the SEXY dream began to become a real possibility.

Far from the big Detroit automakers, Tesla acquired a recently abandoned General Motors and Toyota plant in Freemont California. The original GM plant was opened in 1962 and the previous GM-Toyota joint venture, called New United Motoring Manufacturing Inc., or NUMMI had been established in 1984. The 2009 bankruptcy of GM, following the global recession, saw GM exit the 5.3 million square foot facility. With fortunate timing and an audacious move, Tesla paid just $42 million for the plant once valued at $1bn. Toyota then invested $50 million in Tesla and in addition to the factory, they managed to keep a lot of the plant's equipment for little or nothing. With the new factory, Daimler and Toyota's investment plus Tesla's 2010 IPO, the coffers were loaded to fit out Freemont and begin mass-producing the Model S. The IPO raised $226m. In April 2018, I was fortunate to travel to Freemont on a private tour of the facility. (Due to confidentiality agreements, I cannot disclose more than is publicly available on what I saw inside). The first sight that greets the visitor is the giant red on white TESLA letters on the side of the building. The reception area is basic, with no fancy corporate entrance frills. Factory floor workers and visitors share the same entrance, and it is far from glamorous. There is very much still a sense of a start-up finding its way with frugality evident. With a rudimentary self-service check-in and no mobile phones allowed on the factory floor, I was driven around the vast complex in an electric golf cart. The staff were extremely helpful and polite as they stopped at key assembly line locations and explained what was going on.

Red painted robots are everywhere but there were also a lot of humans doing manual work. Musk is a robot fan and set out to heavily automate as much as possible of the assembly line. After well-publicised issues with the robots, he acknowledged publicly that 'humans were useful'. I witnessed these biological human machines performing assembly tasks next to giant fast-moving red robot arms. One example was when the golf cart stopped at an enormous metal printing press. The tour guy said it was two storeys high and they had to take out a floor to fit it in. In another example of a ballsy start-up move, Musk had acquired the massive metal stamping machine from Schuler SMG hydraulic for just \$6 when it was worth \$50million. Van-sized rolls of aluminium were fed into the press and in seconds 5 door panels would emerge. These were then whisked off to awaiting humans in the next bay for checking and sanding. Human hands were more accurate than the machines for sanding, the tour guide explained.

Musk has a reputation for being tough on his workers and senior team. On my Freemont tour, this didn't come across at all. Indeed, the workers demonstrated a near North Korean-like following for their leader and the Tesla vision. What I noticed in the various presentations and meetings I attended was the passion and singular focus on 'Elon's' grand plan. 'Elon this' and 'Elon that' was almost every second sentence. There is something unique about the team. Although Tesla has an extremely high staff turnover, they are inventive and soaked with a can-do attitude. Having been involved with some start-ups in my past and working in corporate life for over 20 years it is rare to find such a universal belief in the mission of a company.

Freemont was an incredible experience to witness first-hand, however, my visit came just before the period that became known as 'production hell'. Production hell was when doubts rose about Tesla's ability to produce enough cars to meet investor promises. Most of his team did not consult anybody from Detroit when they were building their factory, instead, they set out to create an EV mass production line from the ground up. The creation of an electric car is massively different to how ICE cars are built. With the skateboard configuration of a battery pack, motors and wheels there are no exhaust systems, no fuel systems or pumps and all the usual components associated with normal car production. This meant that the Tesla team had to reinvent, or indeed create from scratch, new ways to produce components and build cars. Need a tool, build a tool was the motto. They were not building a car plant, in the traditional sense, they were building an EV plant, and this had never been done before. Freemont had multiple production issues and delays combined with rumours of poor treatment of staff in the press. Long before the Tesla model, 3 production line could open in earnest at Freemont the team were struggling to keep costs down and deliver on the Model S promises. Although the order-book had swelled for the S, keeping the car below $100,000 was difficult. The stress began to build as cash began to dry up and production delays increased. Tesla came close to bankruptcy on multiple locations during this period and its stock fluctuated widely.

Although assembly took place in Freemont, the global supply chain for electric vehicles didn't exist. Partnerships with Panasonic for battery production were essential to getting the Model S on the road. Tesla did use components from other cars, for example when I had that Mercedes on loan from Tesla, I noticed that the gear selector was the same as the one in the Model S. When Tesla unveiled the Model 3 in 2016 it marked a new chapter for Freemont. Producing the luxury Model S and X would take backstage to the planned Model 3 production. Musk set a target to produce 5,000 cars per week – a gargantuan goal for the still fledging factory.

PRODUCTION HELL

Customers at first overlooked the numerous quality issues with the Model S and X, so smitten were they with their new electric ride. One of the most publicised issues were body panel gaps. Unlike German luxury automakers, that have had over 100 years to perfect their craft, Tesla was rushing the S and X. As they scrambled to scale up production, quality suffered. The Tesla Model S is in the luxury car segment and clear panel gaps could be seen in both the interior and exterior. Leaks in the light fittings and rattling noises in the suspension were frequent. This sort of quality would never be tolerated with an $80,000 - $100,000 luxury internal combustion car. The electric door handles were also problematic as were the windows and occasionally the touchscreen. The evangelistic Tesla followers were prepared to forgive these quality issues to get their hands on the futuristic machines. Tesla got away with these issues for a while at least, but in recent years the quality has come back to haunt them. In scaling up production to handle the massively increased number of daily units for the Model 3, Tesla again overlooked quality in pursuit of production numbers.

Resolving these production issues via Tesla servicing became even worse with long delays in repair appointments. In 2017 Tesla's share price suffered as confidence in the automaker's ability to scale up to become a mass-production car maker was questioned. Doubts over Elon Musk's ability to deliver on his promises begin to grow. Tesla was still not profitable and access to funding begin to become tighter. Rumours of the company's demise began to circulate in the market in 2017. Takeover gossip began to grow, with Apple even cited as a potential buyer. The period of production hell had commenced, and it became a battle for the survival of Tesla and Elon Musk. CEOs and executives of traditional OEMs looked on and often lambasted Musk and Tesla. Several made public comments about the unlikelihood of Tesla ever achieving its ambitious goals. As if by karma, in the years since, many of those manufacturers, who have now chosen to go to electric, have struggled to deliver on their electric production quotas. Indeed, in February 2021 Mercedes announced a one-year delay on their new electric SUV, the EQC, in the American market. This follows Porsche who had serious delays producing the Porsche Taycan electric sports car. It seems that the troubles of Tesla in building a scaled-up EV car plant also plagued the experienced OEMs, proving that electric vehicles are nothing like internal combustion to build. Tesla had to establish the EV supply chain on their own without the experience of

100 years of automotive manufacturing.

Musk knew this was a fight for survival. He took the challenge on himself and personally took control of the Model 3 production hell. Sleeping in the factory and working 24/7 Musk and the team focused on solving the problem. He even slept on the floor of the assembly plant because he wanted to suffer more than the employees dealing with production hell. “The reason I slept on the floor was not because I couldn't go across the road and be at a hotel. It was because I wanted my circumstances to be worse than anyone else at the company. Whenever they felt pain, I wanted mine to be worse.” Musk told a Bloomberg reporter in July 2018. Some staff and management felt that production hell was a step too far while others felt it was a necessary transition to grow the company. In 2017 and 2018 Tesla managed to increase its weekly output of Model 3s from just a few hundred to several thousand per week. Some robots were retired in favour of humans. Large tents were erected in the car parks and surrounding areas of the production plant to handle the overflow. Critics were even claiming that the Model 3 was hand-built, a claim denied by Tesla. Some Wall Street pundits even went as far as to claim that Tesla had been misleading shareholders. These were dark days for the SEXY dream and a period that could have seen a DeLorean style collapse as investors began to question the business model. Several whistle-blower workers released photos of claimed poor quality practices in the outdoor tent known as GA4.

However, the efforts to end production hell began to pay off as deliveries went from just over 8,000 in Q1 2018 to over 63,000 in Q4 2018. In Q2 2019 Tesla beat investor analyst estimates with over 77,000 Model 3 deliveries. Consumer Reports rating the quality of cars placed the Tesla Model 3 at the lower end. Inspection teams were formed to mitigate some of the quality issues customers were experiencing. None of these problems dampened consumer demand with continuing increases in sales and deliveries for the Tesla Model 3 bringing the company to an eventual quarterly profit. Musk was ruthless on performance during this period. Tesla fired hundreds of employees for performance reasons, some justified and some apparently on Musk's whim when he'd become angry at something he may have discovered at the factory, described by one manager as Elon's rage firings. Hitting 5,000 Model 3s per week and beating Wall Street profitability estimates later in 2019 vindicated the alleged near brutal work conditions. Tesla was within single-digit weeks from going bankrupt, according to Musk. Resources were also diverted from the solar business to help with production hell, this came at the expense of market share in solar. The Model 3 was a bet-the-company decision and they pulled it off. Replicating the success was now required to take Tesla truly global. From Shanghai to Berlin and Texas to India the electric car maker is maturing and it's not just about assembling the cars, producing the batteries on a massive scale is vital for success.

GIGAFACTORIES

To achieve global dominance, Tesla created the 'Gigafactory' concept. It's more of a name than a concept, a made-up word combining the Greek word for giant and factory, it's essentially a big factory. Located on a site three times the size of Central Park, work on the first Gigafactory commenced in mid-2014 outside Sparks, Nevada. I visited the $5 billion Gigafactory in 2018. We drove from Reno to the factory and unlike Freemont, which is nestled in the heavily urban sprawl southeast of the San Francisco Bay area, Giga Nevada is in the middle of nowhere. During the long drive to the facility, you gradually witness civilisation disappear. Down 'electric avenue' on a flat rocky desert plain, the vast shiny white complex stands. Wild horses can be seen roaming the grassy outskirts of the complex. Tesla states its "mission is to accelerate the world's transition to sustainable energy through increasingly affordable electric vehicles and energy products" and these are the words that greet you on the towering white wall behind the reception area.

I took a selfie outside the factory only to be quickly accosted by security and told to delete the photograph. No images were allowed inside the belly of the secretive white beast. Giga Nevada or Gigafactory 1, is not like Freemont. This is no repurposed 1960s legacy auto plant, it's a purpose-built, state of the art facility. Inside I was treated to the full tour where battery cell packs and drive train components are built.

Tesla decided to build the Gigafactory to achieve 'vertical integration' - a strategy whereby a manufacturer owns its supply chain and is not dependent on external companies for key components. This was out of necessity as Tesla's production goals require millions of battery packs with tens of millions of battery cells. The factory was conceived to be energy self-sufficient with the entire roof to be covered in solar panels and supported by wind turbines. The state of Nevada awarded $1.3 billion in tax incentives to Tesla as the plant promised to create over 10,000 jobs.

The planned scale for the site is staggering. When fully completed it will have the largest footprint in the world, and second only to Boeing's Everett factory by volume. Giga 1 was essential to the scale-up and mass production of the Model 3. The factory started production of Tesla Power-packs and Power-walls (battery storage for home and businesses) in early 2016. The building is aligned with true north to allow the optimal installation of solar panels on the roof complete with track equipment by GPS. As the Gigafactory has ramped up, the cost of battery packs has reduced accordingly. Reducing the cost of battery production per kilowatt of energy produced is essential for the future mass affordability of electric cars (I'll explain this later). The Gigafactory aims to produce more battery capacity than all the battery capacity of all the current factories around the world. There is a palpable level of intensity inside the Gigafactory – efficiency and high pace activity seem more evident to the untrained eye than at Freemont. The factory floors are heavily automated, red, and yellow robots' rule and humans are sparser than at Freemont. Clutching a cup of Gigafactory 1 Blend coffee, I passed a high-speed feeder belt with thousands of Panasonics 2170 cells racing towards their next stop in the production line. Witnessing the 7,000 plus cells it takes to build a battery pack first-hand, together with the battery cooling systems and overlying battery connectivity, the achievements of Tarpenning and Eberhard truly dawned on me. Without their first experiments, none of this would have happened.

The partnership between Tesla and Panasonic has at times been strained with Musk publicly blaming Panasonic for delaying Model

3 production due to slower than promised battery production at the plant. In 2018 Giga 1 hit over 20 GWh of battery production making it the highest-volume in the world. What started in the Nevadan desert has since been replicated. Musk has stated that 12 Gigafactories are planned worldwide, also stating that 100 could power the whole world. Five Gigafactories plus Freemont (called an assembly plant) form part of Tesla's new worldwide production capacity. Giga Shanghai went operational in November 2019. New Berlin and Texas facilities are both under construction while Giga New York began operations in 2017. Another facility in Tilburg Netherlands completes some final assembly line activities for European models and doesn't get the title of Gigafactory. I have visited Tilburg numerous times and can confirm, with no disrespect to the team there, it is minuscule in comparison.

SHANGHI BERLIN AUSTIN INDIA

Tesla needs global facilities and fast. Tesla may have had a 10-year electric head start on the ICE makers, however, could fall victim to its success. Tesla has woken these sleeping giants. Almost every major manufacturer has announced their intention to go electric in recent years, terrified to be left behind in the electric revolution. In the next chapter, I will cover who is now doing what to make the electric shift and what it means for their legacy businesses. The words 'Tesla Killer' has been uttered at almost every new electric model announced by the likes of Jaguar, VW, Audi and Porsche. It's not the cars that worry Tesla, it's its capacity to compete. As the electric revolution shifts from linear to exponential Tesla will need to produce VW scale numbers to compete. Building new Gigafactories quickly in far-away markets is vital for Tesla not to be out disrupted. To this end, China became the first Giga announcement outside of the US. In July 2018 agreement was reached for Tesla's third Gigafactory, in Shanghai. At lightning speed, the site went from greenfield to producing the Tesla Model 3, employing over 2,000 workers, in just 10 months. Built with an investment of over $2bn, a significant proportion made up of government-supplied working capital loans and a 60-month tax break, the plant aims to produce at least 250,000 cars per year in phase 1 and up to half a million when fully scaled up.

In February 2021 Tesla gained approval from the Chinese Authorities to expand the facility to produce the planned entry-level $25k car – referred to as the 'Model 2'. A design centre has also been established in Shanghai Giga, the first outside the US to design future models for global consumption. With improved production margins and the breakneck development speed, Tesla China will soon surpass Freemont. With Asia now bagged, it was time for the Tesla team to select a new Gigafactory in Europe. After months of rumour and speculation, a greenfield site east of Berlin was selected. Much to the embarrassment of the lethargic German Automakers, Chancellor Angela Merkel offered Musk 'whatever it takes to make the Berlin site a success. At the pace Tesla established China, making the established German brands look pedestrian by comparison, work commenced on the new Berlin-Brandenburg Giga in early 2020. Announcing that the factory would ultimately become the most advanced high volume EV production site in the world, the Germans beat steep competition from other EU neighbours such as France, Portugal, and the Czech Republic. Giga 4 will build batteries, battery packs and assembly of the Model 3 and Y.

Austin Texas became the choice for Gigafactory 5 and in an ironic twist from big oil to zero-emissions mobility the city of Tulsa, Oklahoma tried to lure Tesla to locate its next Gigafactory thereby modifying a giant statue dedicated to the oilmen of the state to Musk's likeness. The 6th Gigafactory may be in Karnataka India. India, now the fourth-largest auto market, ahead of Germany, could build a more cost-effective car, specifically for the lower end market segments. Even though in 2021 Tesla was the most valuable car brand in the world, it is not even in the top ten for global car sales. Toyota, Volkswagen and Ford make up 25% of the global market alone. Toyota, with its legendary lean manufacturing systems, can produce over 10 million cars per year. Going from Freemont + Giga Nevada to four new Giga facilities in just three years increases Tesla's capacity to at least one million cars per year. This would represent just 1.4% of the more than 70 million cars produced globally per year. Tesla has just 1.5% of the auto market sales in the US and less than 1% in the rest of the world. Even with the impressive pace of new factories springing up, Tesla can't achieve an

EV revolution alone, competitors will need to make the electric shift on a massive scale, more on this later in the book.

TRUCKS & CYBERTRUCKS

75% of global freight transport is by road. Road haulage transportation is growing worldwide and is responsible for around one-quarter of all vehicle CO_2 emissions. This is a problem and market Tesla could not ignore. On top of this, international regulations are setting CO_2 emissions standards for the years ahead. In the EU, regulation 2019/1242 aims to reduce 54 million tonnes of CO_2 emissions, under the Paris Agreement from road freight. A system of credits and incentives for lower and zero-emission freight vehicles is being introduced first, to be followed by future financial penalties in the case of non-compliance from 2025.

In true Tesla style, the Semi Truck team set out to radically change the design of the traditional cab. I was treated to a presentation during my trip to Freemont by members of the Semi team. They had done their homework. Measured gross tonnage to Khw of power required in a multitude of environmental settings was thoroughly researched and presented. They also involved experienced truck drivers in the cab's design and analysed typical freight routes, truck stops and driver behaviours. The result was the Tesla Semi Truck, unveiled in November 2017. Known as a Class-8 semi (a truck with a gross vehicle weight exceeding 33,000 pounds / 15,000 kg) the design was again revolutionary. The driver sits in the middle of the cab facing two horizontal touchscreen displays. Tesla began testing prototypes in early 2018 to haul battery packs from Nevada to Freemont.

Tesla had secured over 2,000 orders as of early 2018 and major brands signed up for starter orders like FedEx, UPS and Walmart. Using four Tesla 3 motors and a 500-kWh battery pack the Semi will be capable of 0-60mph in 5 seconds unloaded. Of course, it's not meant to be a sports vehicle and what matters to truckers is pulling power. Over 33,000 individual, new design, 2170 battery cells (so named due it its 21mm width and 70mm length) will be in the Semi power packs providing 800 horsepower. With promised autopilot features, as with

SEXY car models, Tesla aims to ultimately build fully autonomous trucks to improve efficiency, safety and delivery times. As of early 2021, the truck is almost two years behind schedule and unlikely to be mass-produced until after Tesla begins making its own batteries. Tesla has competition with existing manufacturers and start-ups moving into the electric truck space. Some don't believe that batteries are the way forward, with the heavyweight to load ratios. Some, like troubled start-up Nikola, are trialling hydrogen fuel cells (more about the hydrogen hype later). Big players like Toyota, Volvo and Volkswagen also want to get in on the semi-truck electric action. Tesla is advancing their supercharger network to become mega chargers capable of charging these giant battery packs in 30 minutes or less. It is likely, and rumours have it, that Tesla will also begin developing smaller so-called last-mile delivery vans in the future. I will explain in the next chapter how this has become a huge market interest area with multiple start-ups moving on the van market.

In another sector, going all out on a radical new design, Tesla unveiled the CyberTruck in November 2019. In many US states, the pick-up is king. Tesla aims to break this market with something completely different to what has come before. The Ford F150 is a standard-bearer in the pickup world. Musk unveiled a futuristic dystopian style machine claiming to be bulletproof; he had the lead designer, von Holzhausen, try to smash its windows with a steel ball live on stage. This didn't quite go to plan as the window wasn't fully closed during the demo, so duly shattered! Nevertheless, the CyberTruck has turned heads. It is a giant machine and plans to have pull power above even the biggest pick-ups. As a challenge/insult to Ford, a video surfaced with a Cyber truck pulling a Ford F150 backwards as it tried to accelerate away like prey caught in the jaws of a predator. Musk said the design was inspired by the sci-fi movie Blade Runner and the Lotus Esprit driven by James Bond in The Spy Who Loved Me. The truck's body is made from DeLorean-like stainless steel. Every announcement from Tesla now comes with extreme public and media scrutiny. Such was the radical design of the CyberTruck that Tesla share price dropped by almost 6% following the announcement. The market is estimated to be worth over $65 billion annually with the US accounting for 60%. Pick-up trucks represent around 4% of the worldwide total auto market and growing. Once again Tesla has competition. New entrants like Rivian have produced a more traditional styled electric pickup truck. With over 250,000 pre-orders in the first three months, time will tell if Tesla's extreme radical design is a winner or whether customers stick to the more traditional format.

$25K CAR

The holy grail for mainstream electric car adoption is an affordable mass-market car, $25,000 or less. As in the past with Ford's Model T, the VW Beetle, and Citroen's 2CV, breaking this market will make electric the default choice for all. The key to producing an affordable mass-produced electric car is battery cost. Energy output to battery cost ratio has consistently improved for the last 10 years and will continue to drop (more on this later). Musk announced in late 2020 that Tesla is to begin working on the 'possibly lower than $25K' car during Tesla's 'Battery Day Event'. Dubbed the Model 2, if successful, it would effectively accelerate the end of the ICE age even sooner.

Volkswagen launched its ID3 in 2020. The VW Golf styled car, designed for the masses, is still relatively expensive at around $40,000, although Volkswagen promise cheaper models are coming. Whoever gets to this holy grail of $25,000 or less will no doubt secure a huge market share and quickly. Given Tesla's massive brand appeal and loyalty, combined with newly scaled-up Gigafactories worldwide and their EV head start, their potential to dominate is huge. As with the new promised Apple iPhones, the rumour mill has gone wild with concept designs for the Model 2. It is unlikely Tesla will radically deviate from the Model 3 style because the public, in the main, likes boring when it comes to the smaller everyday car. The Volkswagen ID looks almost exactly like a Golf, which hasn't changed in design for over 20 years. The story of Tesla is much more than just the actual physical electric car. Tesla has developed, across the globe, its Supercharging network. This unique advantage combined with 10 years of software development puts Tesla in a position to deliver more than cars, they have the potential to be the global EV winner for a very long time. Software and data are their real competitive advantage.

BEYOND THE CAR

One of the reasons for Tesla's massive stock market valuation is the belief in its potential to generate future revenues beyond just selling cars. With up to eight cameras, always-on internet and GPS, Tesla has been acquiring mammoth amounts of data to develop its autonomous capabilities. With real-time feeds from every Tesla car on the road, the company has been mapping and learning fully autonomous software capabilities. Known as Autopilot – this could potentially give Tesla a worldwide lead in the expected future fully self-driving car. Tesla cars already have Autopilot, Summon Mode and Full Self Driving (FSD). There are six stages for autonomous cars - I'll cover them in the next chapter, however, safe to say that for now, Tesla is way ahead, not necessarily because of their technology, but because of their live data gathering head start. The Tesla Model S has been on and off the fastest accelerating production car in the world, occasionally losing its title to others like the electric Porsche Taycan. Borrowing from the nerdy Space Balls movie Tesla has named its fastest Model S modes as 'Ludicrous' and 'Plaid' with 0-60 mph in just 1.9 seconds. While these speeds are exhilarating, they are of little daily use to the average driver. The purpose of these modes it so to demonstrate the innovation and efficiency of Tesla as a leader.

The next 'beyond the car' potential is the charging network. From the early days, Tesla began installing Superchargers on major routes across the US and Europe. The first Supercharger opened in September 2012 and as of 2021, there are over 2,500 stations with over 24,000 individual chargers. These Superchargers are high-capacity electric direct current units, capable of charging a Tesla Model S from 0 to 80% in 25 minutes; I dive into how Supercharging works, how it's a massive advantage to Tesla and my experience driving all over Europe and the US. The biggest takeaway from the Supercharging experience is EASE OF USE. The car + driver + charger is 'connected' via the Tesla touchscreen for an easy drive-plug-drive, with no buttons, screens, or options. Unlike other car manufacturers, Tesla has its charging infrastructure, and the competitive advantage and user experience cannot be overstated. It's also a huge potential future revenue generator by selectively allowing other manufacturers to use the network at some point.

The next is shared mobility. The Tesla Model 3 comes with, a yet to be activated, inside camera facing the occupants. This is a feature for a planned peer-to-peer, Airbnb-style, sharing service whereby owners could short term 'rent' out their vehicles to others. Musk has also stated that as self-driving technology develops the Tesla Model 3 and Y could be 'Robotaxis' with the camera protecting against vandalism. Being one of the safest cars in the world combined with partial and full autonomous features, insurance becomes another area to examine. With its lower probability for an accident, real-time driver behaviour tracking and new pay per use insurance models, Tesla launched its insurance offering aiming to deliver drivers 20%-30% lower premium costs with an 'Autonomous Vehicle Protection Package'. Electric cars combined with preventative accident technologies and future autonomous cars may redefine how traditional car insurance has worked for a century, another huge area for potential disruptive leadership by Tesla. With the controversial acquisition of Solar City, Tesla can offer a completely sustainable energy package to consumers. Solar tile roofs (instead of unsightly solar panels) offer a near-invisible solar solution. A Tesla Powerwall (effectively a car battery on the wall) in your home, powered by the Tesla roof can then provide energy independence for your home and car. Car to grid systems in some geographies can allow your car and roofs excess energy to be sold back into the local energy provider. Of course, not all locations enjoy the sunshine levels of Texas or California, however, solar technology is improving and the cost to energy ratios are reducing, making this technology closer to becoming mainstream in the years ahead - a lot more on this later. Tesla, therefore, can offer a one-stop-shop from energy to mobility.

Tesla doesn't just provide power walls for homes. Its energy solutions division has made high profile power solutions to solve energy crises in places like Puerto Rico and Australia. After Puerto Rico was devastated by Hurricane Maria, Musk said he could potentially rebuild its entire electrical infrastructure with Tesla solar and Tesla battery packs. In

2017 Tesla installed power packs and solar panels at a major children's hospital on the island. Musk himself donated $250,000 to the efforts. In Australia, Tesla has supplied massive battery installations to ease pressure on Victoria's energy grid. Following a 2016 storm, South Australia experienced a total blackout after, 80,000 lightning strikes and two tornadoes. At the Hornsdale Power Reserve, there is now enough Tesla battery capacity to store up to 100 MW of energy. This gives the region a buffer against future outages. Musk made a bet that in just 100 days the energy farm could be up and running and if they missed the deadline, it would be free. Musk won the bet.

Advancing battery technology (explained later in the book) and Tesla's new energy storage credibility gives the company another big head-start over competitors and offers potential huge future revenue streams. From self-driving cars, car sharing, solar energy, large-scale energy storage, insurance to its Supercharger network, Tesla has built an impressive conglomerate in the future of energy and mobility. All of this has been achieved mostly only in the last decade. How can a company achieve so much in a relatively short period? The answer is complex. Partly timing, partly vision, partly the team, partly the complacent incumbents and hugely the Elon Musk Effect.

THE ELON EFFECT

Like with many revolutionary companies, such as Apple, it often seems like the company cannot continue without their visionary leader. Indeed, it was hard to believe that Apple could become even more successful and bigger without Steve Jobs, however, as of 2021, Apple is six times the size it was when Jobs died a decade earlier. Great companies and products outlive their inventors and endure. Musk will leave Tesla someday and it will be up to the team and successor to drive it to new heights. Musk came close to leaving already, particularly when back in August 2018 he issued a Tweet that he had 'funding secured' a $420 price tag per share to take Tesla private. Seen as potential market manipulation, this Tweet landed Musk in trouble with the Securities and Exchange Commission (SEC).

Following the Tweet and some suggestion that perhaps Musk was becoming unhinged, the value of Tesla dropped over 38%. The SEC stated that Musk had issued a "false and misleading statement. Fighting back and publicly challenging the SEC he ended up paying a $20 million fine with the company paying a further $20 million fine. Almost childishly, Musk again took to Twitter to mock the SEC, referring to them as the "Short seller Enrichment Commission" in the days after the settlement. Musk's reference to the army of short-sellers, non-believers betting that the stock will drop, came after years of battle between the visionary and Wall Street gamblers. These short sellers were heavily burned as Tesla's stock price reached new heights in 2021. Other controversies followed for Musk. When on the Joe Rogan show, he very briefly smoked some marijuana live on air, drawing heavy media criticism and another stock drop in the days that followed. Musk did speak on the show for hours about the future of humanity's survival but the few second of weed is largely all that is remembered from the interview. Musk landed in more hot water when he tweeted the words 'Pedo Guy', meaning Paedophile, referring to a cave rescuer rebuffing his offer to save trapped schoolchildren in Thailand - resulting in a court case, which Musk won.

He is the CEO of a now huge company; that comes with responsibilities and certain conservative expectations. Following his Tweeting and other public events, a board overseer was appointed to keep an eye on him in the form of a new chairperson. Very often companies outgrow people and people outgrow companies. Musk is special, nerd like and often awkward in public situations. He is no doubt extremely talented and extremely intelligent and genuinely believes his vision to become a multi-planetary species. The Iron Man movies parallel Musk's eccentric mad genius. Robert Downey Jr was given a personal tour by Musk of the Space X facilities in 2007 for which Musk had a cameo in the 2010 movie. He enjoys a near Messiah-like following on social media. Doubters and critics of Tesla are often berated by his loyal following and he makes grandiose claims and sets unachievable deadlines. However, it is these grandiose claims and deadlines that make action happen.

Viewing Tesla's mission as a project, there are only three success factors in every project - time, quality and cost. Getting all three right rarely happens. Musk has delivered on his grand plan and in mid-2016 he set out the second grand plan – Master Tesla Plan, Part Deux as follows.

• Create stunning solar roofs with seamlessly integrated battery storage
• Expand the electric vehicle product line to address all major segments
• Develop a self-driving capability that is 10X safer than manual via massive fleet learning
• Enable your car to make money for you when you aren't using it

One or all of time, cost and quality may suffer in achieving the Part Deux project, however, I would bet it will get done, and probably more if the momentum of the past can be replicated. From its founding in 2003 Tesla has achieved phenomenal disruptive change. The pace of this change has accelerated and finally pushed other sleeping car manufactures into action. History will record Tesla as the instigator at the end of the destructive Big Oil and ICE age, tipping the balance to the electric age.

CHAPTER 5

TIPPING POINT

BACK TO THE FUTURE

Tesla brought the electric car back to life, making a zero-emissions mobility future a real possibility. Electric cars still represent just a tiny proportion of cars on the world's roads today, having a minuscule impact on CO_2 emissions so far. That's all going to change and fast. Over the past ten years, I have witnessed many statements by politicians, read reports by research groups and listened to the CEOs of legacy carmakers talk about a linear, gradual switch to electric cars. The 2040s and 2050s were commonly claimed to be the time frames when electric cars overtake ICE cars. These predictions were made, at best, in ignorance or, at worst, in 'survival mode' self-interest. History has repeatedly shown that once a new, and better, technology gains traction it often starts slow but then quickly goes exponential. We have reached that point in the electric car transition, the tipping point. Every indicator now points to the rapid adoption of electric cars. This is accelerating because of the climate imperative, consumer demand, government bans and now, finally, the OEMs waking up to the electric reality.

This is the third time lucky for the electric car. From 1890 to 1910 the electric car almost made it before being wiped out by the gasoline engine. In the early 1990s, the EV almost made a comeback, only to be swiftly wiped out again. This time it's different. This time there is no going back. The Big Oil ICE Age will end, faster than most people think. I predict that by 2025 EVs will dominate and few in developed economies will be purchasing an ICE car. The resale values of ICE cars, starting with diesel, will begin to collapse from 2024. The world's major fleet owners, like rental companies, government fleets, corporates, and industry (who typically replace their cars in three-year cycles) will realise that replacing ICE cars with new ICE cars in 2024 would be financial suicide. A 2024 ICE car will be financially obsolete by 2026, its resale value will be near scrap as EVs go mainstream. EVs have already crossed the point where the total cost of ownership versus an ICE is better. By 2026, with ever reducing battery pack costs, a deluge of new affordable models and range anxiety all but banished as a myth, buying an ICE car will make zero sense.

Before we go forward, we need to get closure on the past. Thomas Edison's first car was electric, a Baker. The 1909 Baker Electric car was a revolutionary car for its time: simple, low maintenance and with a range of over 100 miles. Scary steam-machines did not appeal to women at the turn of the last century and the Baker was fitted out with a luxurious, feminine interior specifically marketed at women. Henry Ford's wife owned an electric car. The first White House car was a Baker, used by President Taft's wife, Helen. Over 15,000 6volt Bakers roamed the streets, supported by on-street charging stations. They were run on Thomas Edison alkaline batteries that could be used over and over by just occasionally cleaning and adding water. Like a Tesla today, the Baker was designed with brutal simplicity and ease of use. No hand cranking, unlike an ICE car of that time, no oils, fluids and noisy dirty engines. A single 'tiller' style leaver was all that was required for acceleration - reaching 23 miles per hour. Although it was a lot slower compared to the Model T's 42 miles per hour, it was designed as a city car. After losing control in a mile a minute speed test, a souped-up Baker crashed into a crowd of spectators in 1902, killing two and damaging the image of electric cars.

It wasn't just the accident that killed the electric car at that time. Politics and money played their part. In the US the Electric Vehicle Company was once the country's biggest automaker. In addition to selling them, they also supplied their cars on a rental basis. Cars were rented per day and returned for charging at night. These cars were on-demand EVs, pay per use with charging included. It has taken 120 years for this model of 'Mobility-as-a-Service' to return. Electric, shared, now contactless and soon autonomous is the future of urban mobility. A 1901 loan controversy scandal bankrupted the Electric Vehicle Company. Baker, Studebaker and others all either disappeared or dropped their electric cars as ICE took over. The 1912 invention of the electric starter eliminated cranking ICE cars every time you wanted to move. This innovation combined with better-paved country roads and gas stations, allowing for longer journeys out of the cities, doomed the EV. It was another 80 years before a serious resurrection looked possible.

In 1990 the State of California enacted the Zero Emissions Mandate. This new regulation set a production target of 2% of cars to be produced by automakers to be zero-emission by 1998, increasing to 10% by 2003. In response, General Motors produced the EV1 in 1996. With a 70–100-mile range and a maximum speed of 80 miles per hour, the odd-looking but futuristic machine gained popularity when celebrities such as Tom Hanks and Danny DeVito bought and lauded praise on them. Other manufacturers began retrofitting ICE models with batteries so to be compliant with the new law. Whilst there appeared to be baby steps towards electrification, in the background the automakers were fighting and lobbying to have the law scrapped. Between Chrysler, Toyota, Nissan, Ford and GM, 5,000 EV's were produced. As with the tobacco companies in the past, major automakers spent millions on lobbyists and lawyers to have the law repealed. In the 1990s the SUV began to become extremely popular. These gas-guzzling monsters produced high-profit margins for the OEMs, and they tried everything to get rid of the forced EV production targets. Their anti-EV campaigns coincided with the development of new SUV models and monsters like the Hummer – like drug pushers upping the potency of their poisons to keep their junkies hooked.

The campaign was highly orchestrated; for example, the Western States Petroleum Association financed lobbying to prevent utility companies from building public charging stations. Demonstrating Jekyll and Hyde behaviour, GM even ran Super-Bowl advertising for their EV, whilst at the same time running negative campaigns highlighting the technology deficiencies with electric cars, most of which were not even present in their EV1. In 2002, then president, George W. Bush, joined the auto industry in a suit against California. With relentless pressure on the California Air Resources Board, eventually, the auto makers' and Big Oil's persistence won out. The law was reversed. GM recalled all their EV1s and crushed them. This public, book burning-like display, took place despite EV1 owners pleading to keep their cars, even forming a protest outside GM's premises. Despite some owners pledging $1.9m to pay the residual value on their EV1, GM ignored the offer, recalled the cars and destroyed them.

In 2006 the documentary "WHO KILLED THE ELECTRIC CAR" ironically starred Tom Hanks, as the documentary chronicled how big oil and automakers colluded to wipe out the EV. The auto industry didn't stop there. For years after they continued to keep the electric car down. GM teamed up with the Trump administration to successfully roll back President Obama era emissions standards. The second coming of the EV was over, confined to history by the victorious oil and ICE industries.

Tesla resurrected the EV and grew from being an annoying insect to the auto-industry into a disruptor that has shaken it to its core – the EV is back for the third time and this time it will be different...

THIS TIME IT'S DIFFERENT

During his presidential campaign, Joe Biden said "we have to take combustion engine vehicles off the road as rapidly as we can". Is it the third time lucky for the EV? Yes, however, even with Tesla's initial success, the future of electric dominance looked fragile with Tesla's woes in 2017 and 2018 and the rest of the industry still clinging on to ICE. Having killed EVs, GM made a very public about-face in February

2021 with its announcement to have 30 electric models by 2035. Comedic actor Will Ferrell starred in a Super Bowl GM EV commercial that when viral. "Norway is out EV-ing us!" the ad proclaimed, stating GM's intention to establish the US as the top electric nation in the years ahead. There is intent now, however, it is very recent. Not long before, GM had found a friend in Donald Trump. Just six months before the Will Ferrell ad GM was fighting alongside Trump to curb emission standards from the Obama era. The auto industry is the reason behind the unnecessary prolonging of the ICE age and is jointly guilty with the oil industry for our climate nightmare today.

With Biden's election and the immediate reinstatement of Obama era emission standards, the likes of GM had no choice but to race to make the electric transition. For the US to catch up with Norway is a different matter though. The ad is hilarious, however, Norway is a special case. A wealthy nation, ironically from its oil past, they were able to make radical policy changes to encourage EVs. In 2012 just 3% of Norwegian cars sold were electric, in 2020 that reached 54.3%. The Tipping point was reached for Norway but how? In summary carrot and stick. Introducing big subsidies for electric cars in Norway shifted affordability away from ICE vehicles. In parallel large scale charging infrastructure and consumer education tipped the balance to EVs. Norway is, in 2021 a world EV leader. However, with a population of just 5.3 million, compared to the US's 330 million, it has a relatively small global impact on CO_2 emissions. For a meaningful change, the biggest car markets, the USA, China and Europe need to do the same as Norway.

The US has a very long way to go to come even close to Norway. Electric cars accounted for just 1% of the 146 million new car sales in the US between 2011 and 2019, according to usafacts.org. It's not just about incentives, it's about attitudes and long-standing motoring beliefs. There remain many myths about the electric car in America. In 2021 I was pitching to a Texan based venture capital firm. The VC partner on the other end of my Zoom call amazed me when he said, "I love your customer experience model but why are you guys experimenting with electric cars?". Despite big US automakers making public announcements about their electric ambitions, large swathes of the US public need EV convincing. To replicate Norway and tip the balance to mostly EV sales, the US will need to regulate and incentivise on a massive scale. With states like California planning future ICE sale bans, this time unlikely to receive the same concerted attack from the ICE industry, the US electric future seems inevitable. However, even with these planned state bans, unless the public can be convinced and the necessary infrastructure is massively rolled out, these dates could get pushed out further.

The end of Trump's reign was likely the last breath of resistance to US electrification. Without political support in Washington, automakers will receive a cold reception to any attempts to block emission standards or to try contravening Paris Agreement targets. Whilst they may make big advertising campaigns claiming to go electric for now at least, their money-makers are ICE cars. This is the dilemma facing the auto industry and placing Tesla just so far ahead. Tesla still has start-up agility, doesn't rely on a legacy product to bring in the cash or have all the historical infrastructure and work practices holding it back. This is a real issue for legacy automakers. If the US adopted a Norwegian style disincentive to buy ICE cars and a strong incentive to buy electric combined with, say a doubling of gas prices, many automakers would likely collapse. Revenues from ICE would stall and they would not yet have the EV production capacity and variety of models to supply demand. For this reason, and the current lack of US charging infrastructure, the EV revolution won't happen overnight in the US just yet. However, what has started slower in the US will likely rapidly accelerate.

The first early realisation that legacy automakers had been asleep at the wheel while Tesla grew around them came from then-GM CEO, (2000 – 2009), Rick Wagoner. Wagoner admitted that GM had "squandered their early lead in electric-hybrid tech". Wagoner claimed that scrapping the EV1 was the worst decision of his tenure. Losing his job, with GM's 2009 bankruptcy, his successor, Dan Akerson stated that Tesla was just a tiny threat to GM. Despite their puny size at that time, he still created a study team, in 2013, to determine just how big a threat Tesla could become stating that "History is littered with big companies that ignored innovation that was coming their way because you didn't know where you could be disrupted,"

Speaking about the success of the Model S, he said "It's fascinating. I don't know if it's going to work or not work. All I know is, if we ignore it and say it's a bunch of laptop batteries, then shame on us." This research group ultimately led to the launch of GM's Chevrolet Spark in 2013 and later the Chevy Bolt in 2017. These early models laid the ground for GM's Feb 2021 proclamation to sell one million EV's by 2025. GM is the first major US automaker to go 'all-in' on electric.

I have focused a lot here on GM, as it's the biggest US automaker with an anti-electric past, however, others have also declared their electric future. This time it is different, this time the world's automakers (not all yet, but many) are declaring the end of their ICE strategy. They have largely given up the anti-electric fight. They are beaten down and finally give in. Reaching closure on its zero-emissions war, GM dropped its lawsuit against emission standards in November 2020 and within a few months underwent an electric 'a coming to Jesus' moment. This time, the third coming of the EV is different, this time it's for real, carmakers are going 'all in' on electric.

GOING ALL IN

The world is reaching a global climate crisis tipping point. Extreme heatwaves in Canada, extreme flooding in Europe and China in 2021 are the signs we were warned about. We must reach zero CO_2 emission by 2050 to avoid total disaster. The public is gradually growing in this realisation and looking for options in their carbon choices. It is not just the climate imperative that has finally driven automakers to go electric, it is also a survival response. In our fast-changing world, the necessity to adapt quickly is as much about survival as it is a success. If they wait another four years, they will be 'Blackberry-ed'. GM has announced 30 new electric models by 2035. It's an ambitious goal for a massive juggernaut like GM, however, a pedestrian pace compared to the rise of Tesla. Having 30 electric models by 2035 would be 32 years after the founding of Tesla and 23 years after the launch of Model S. I believe this is too slow.

The pace of change in the 21st century is accelerating, driven by tech-savvy consumers and rapid exponential technology advancements. For GM their going 'all-in' is more like slowly stepping into a cold swimming pool than a full-on dive. Still, it's a step in the right direction and many others are now planning to follow. The Alliance for Automotive Innovation (AAI), representing giants like GM and Ford has made commitments to work on a national electrification program that would bring all automakers "under a unified set of common requirements". The results of the November 2020 US election tipped the balance in favour of electric cars in auto industry boardrooms. A global sea change is underway.

Ford, Jaguar Land Rover (JLR), Volvo, Daimler and BMW have all declared grand electric ambitions. The announcement in January 2021 that Ford is going all-in on electric, the instigator of mass ICE adoption 120 years ago, was momentous, to say the least. Ford's new strategic goal is to sell only electric cars in Europe by the end of the 2020s decade. The iconic Ford Mustang now has an electric twin. Ford is still building ICE Mustangs, but you can now also buy a Mach-E fully electric car. By 2026 Ford aims to have 100 percent of its passenger road cars all-electric or plug-in hybrid. CEO, Jim Farley stated that Ford will be investing $22bn in electric through to 2025.

In a sign that Wall Street endorses GM and Ford's new strategy both companies stock values increased 100% and 65% respectively in just six months. It's a lofty ambition for Ford when in 2020 it didn't sell a single electric car in Europe. Ford and others do have a legacy holding them back to a huge extent. It's not just the factories but also, it's their mindset. One of the reasons for Tesla's success is that they built a battery pack first then a simple, easy-to-use customer interface. Unfortunately, the legacy carmakers are not doing the same in most cases. I've had this experience on multiple occasions when getting into brand-new EV models from other manufacturers. Yes, they are probably better finished and have a more luxurious interior, for example, the

Jaguar iPace, however, they reek of the ICE age. With buttons, dials, complex charging navigation systems they are, compared to Tesla, almost unusable. Asking a consumer to shift from the long-held habits of internal combustion to electric means it must be easy. It should be even easier than a gas car. To date, and not to single out any of the fine EV machines produced so far, cars like the Audi E-Tron and Jaguar iPace simply were not built with 'start again thinking' and are full of legacy features you find in ICE cars. It won't be easy for these manufacturers to replicate the start-up thinking and agile approach of Tesla.

While Tesla builds Gigafactories in less than a year, existing ICE players already have all this manufacturing capacity and infrastructure in place for ICE. Ford has decided to convert its 90-year-old German factory, in Cologne, to produce a small electric (Ford Fiesta) category car by 2023. Ford will be using Volkswagen's MEB platform (German acronym for modular electric-drive toolkit). Time will tell if a company with the heritage and legacy of Ford truly reinvents itself and produces world-class electric cars to compete with Tesla and the slew of others now going electric.

In response to Tesla and coming out of the ashes of Dieselgate, Volkswagen has also declared its future is electric. The People's Car for everyone, iconic legendary VW has already produced some of their ID range. This includes the Golf-like ID3 and the new, Tesla Model Y-like, ID 4. VW's strategy is to have both electric versions of existing models and brand-new, from the ground up electric cars like the ID. Committing almost $15 billion alone to vehicle battery supplies, they also announced the conversion of 16 of its major factories to build electric cars. By the end of the decade, Volkswagen plans to have almost 300 ICE car models replaced by electric. Like General Motors, Volkswagen, while betting big on electric, only expects to have 40% of its sales electric by 2030. Again, I think this is under ambitious and potentially too slow for Volkswagen. It is a phenomenal pivot in strategy from the days of Dieselgate, less than a decade ago, however possibly not enough for Volkswagen to retain its global dominance in the mid-level car segment. Affordability is the key and battery production unit costs are essential if a company like Volkswagen can truly convert the mass market. Volkswagen's ID3 is simply too expensive to be treated as a true people's car for now. Volkswagen has said that by 2026 they will be finished developing their internal combustion platform and only develop electric. Like Tesla, Volkswagen's luxury brand Porsche released the beautiful Porsche Taycan in 2020. This car has been very much built for Porsche followers, it's expensive, high-performance, and outfitted with every conceivable gadget and extra. In typical German style, the car is probably over-engineered for example with two charging ports, one for fast and one for slow charging. Once mired in the Dieselgate scandal and paying billions in subsequent fines it was quite incredible to see VW CEO Herbert Deiss and Elon Musk together on stage and trading compliments during the 2019 Golden Steering Wheel Awards in Berlin. Who would've thought just a decade ago the unknown electric start-up would be shoulder to shoulder with the auto-greats? The Tesla Model 3 won the award.

In March 2021, VW held its "Power Day" replicating Tesla's battery day held in September 2020. At the event, Volkswagen said they would build six Gigafactory equivalents in Europe for battery production by 2030 and aim to overtake Tesla's sales by 2025. Likewise, it made grand claims for its battery production. Some of these claims included a future move to solid-state batteries and breaking the threshold of the precious $100 per kWh. I will cover more later the revolution in battery production and technology as this is fundamental to the success of mass EV production. Volkswagen is now in a head-to-head race to be the leading electric automaker in the world. They are unashamedly virtually copying Tesla, but this is a good thing to come out of the ashes of the Dieselgate scandal.

The big luxury German brands, Daimler and BMW, are also pointing to an electric future. Not quite the same dramatic announcements of 'going all in', as with others, however, their electric journeys have very much begun. Daimler chairman, Ola Källenius, said automakers face a choice to either get dragged kicking and screaming into the new reality or to become "one of the architects" of a new, cleaner mobility future, adding that "Almost regardless of regulation, we are now on this path," he said in 2020. Mercedes launched the SUV Electric EQC in 2020 with a very similar design to its ICE cousin and plan dozens of new electric models by 2025. The luxury automaker however didn't exactly focus on the tailpipe or emissions during its glitzy 2020 unveiling of its new flagship ICE S Class.

Another iconic carmaker, Jaguar Land Rover (JLR), has stated that by 2025 Jaguar will be all-electric and by 2030 Land Rover will follow. For Jaguar to reinvent itself, it will also be dropping some old brand icons. They already have an electric SUV, the iPace, and indicated they will be considering new models to compete with the Tesla Model 3 and Y. Building smaller cars, electric-only, will be a new departure for Jaguar, however, an interesting one for their brand. Jaguar intends to use the Electric Modular Architecture (EMA) platform for its future electric cars. In early 2020 Mercedes sold 87,000 xEV's - a term which refers to both plug-in hybrids and all-electric vehicles. This represented 7.4% of Mercedes sold in 2020, up already from just 2% in 2019. With the demand for electric cars increasing by 185% in 2020 in Europe, Daimler will likely accelerate its 'great transition'. Together with BMW, they have both been slow to acknowledge the electric revolution and invest in new electric technologies.

After years of electric uncertainty, BMW announced they aim to have nine new electric cars by 2025. Although they introduced the i3 BMW way back in 2012, the Bavarian auto legend isn't quite yet going 'all-in' on electric. BMW are more cautious as they expect different speeds of electric take up from market to market. Indeed, later I will examine and describe how there could be a class divide amongst countries as the world electrifies with poorer countries being left in the ICE age.

Apart from Nissan, with its long-standing Leaf, the Japanese have been similarly slower to embrace electrification. Nissan planned to have eight EVs by the end of 2021 and aims to sell over one million hybrid or electric vehicles per year globally by 2024. Toyota remains a big outlier, a black sheep in electric mobility. Toyota is the king of hybrids, and they have focused on their hybrid strategy – (I'll cover how Toyota have been false advertising their 'self-charging hybrid' later) and haven't made the same transformation commitment as other global players. Toyota also focused on hydrogen fuel cell technology. This is a mistake and again, I'll explain why later. After years of fighting the inevitable, Toyota, in early 2022, announced finally to move seriously to electric. They announced a new dedicated modular battery-electric platform called thee-TNGA, like VW's MEB platform. The world's largest car producer is making what appears as hastily prepared electric announcements. For years, Toyota has been ignoring pure electric cars as a strategy until late 2021. Their luxury brand Lexus's CEO, Akio Toyota, has publicly criticised electric cars. Perhaps they have come to grips with the electric revolution and are truly planning to follow the crowd or they are seriously behind - now in panic mode – threatening their future as a global auto leader by coming very late to the party.

Many others like Citroen, Hyundai/Kia, Renault, Peugeot have all declared their future is electric. The 2021 created Stellantis group, now the world's 4thth largest automaker with 16 brands such Peugeot, Fiat Chrysler, Citroen, Jeep and Opel have declared an electric future, planning to have all its current models in fully electric or hybrid by 2025. Almost all automakers are now turning up to the race that Tesla created and already has an enormous head start in. Making announcements to tool up and build massive electric car building capacity is just one part of sustaining the EV Revolution. The second part is convincing the public to switch electric. For this to happen four key elements need to be solved.

1. Affordability
2. Charging
3. Battery Capacity
4. Dispelling Fact from Fiction

There is a huge number of falsehoods and misconceptions on electric cars I'll delve into in the next chapter.

More has happened in the last ten years than the last 100 in the automotive industry. New electric car makers, many of which you have likely never heard of, have all appeared on the scene with big promises. And it's not just cars; trucks, vans, scooters, mopeds, aeroplanes, ships - almost everything that can move is now on a path to electric.

Fisker, Rivian, Lucid, Nio, BYD, Byton, XPeng Motors, Arrival, Lightspeed, Sono, Aspark Owl, Faraday Future, Lordstown, Polestar, Nikola. These are just some of the new car makers to spring up over the last few years. Never since the first 20 years of the last century have so many new start-ups entered a market. Why? Because the electric transition could be a goldmine. If you believe that it's not going to be linear but an exponential electric transition and that the existing carmakers won't be able to adapt quick enough, then this new world could be an automotive cash cow. This has happened repeatedly when new technologies are introduced. Within a few years of the mobile phone becoming practical for the masses, new manufacturers began to spring up everywhere, the same for the video game consoles, home PC and even fast-food chains – everyone wants a piece of the action. These are real electric companies with serious investment and backing already. Many are introducing their electric cars, vans, trucks and buses onto the market in 2022. Some, like Lucid Motors, are squaring up to directly compete with Tesla. Others are going more mainstream, with mass offerings to compete with the likes of Volkswagen and General Motors. The Chinese (more later in the chapter) are scaling up extraordinarily fast with a host of new electric cars coming on the market.

Before Covid shut down live events, I was lucky to attend the Frankfurt motor show at the end of 2019. I was invited to speak at the show on eMobility. Spending two days walking around the massive exhibition and looking at all the major car brand displays, it became obvious to me which car makers were truly embracing the electric revolution and which were just making token gestures. For example, at that time, Ford had zero to present on electric but still mentioned eMobility on their displays while others like Toyota didn't even mention electric at all. The crowds' attention was on the new electric upstarts. There were queues to sit in the new Chinese BYD and a frenzy of activity around Porsche's new electric Taycan. Not all the new electric car companies I mentioned above were at the show, but those that were there certainly stole the show. The old players just looked like old news around these futuristic electric machines. I'm not going to cover every new electric car model coming on the market, but I am going to mention a few that are hitting the headlines.

Back in 2004, I watched Ewan McGregor and Charley Boorman's epic motorcycle adventure from London to New York, the Long Way Round. In 2019 they repeated this epic adventure, the long way up, from Ushuaia, in Argentina, to Los Angeles. This time they did it on electric Harley Davison prototype motorcycles. Perhaps more interestingly, their journey was supported by two electric prototype pick-up trucks from start-up Rivian. Although the program cast electric in a positive light, there were some issues of course with charging in remote regions. Rivian is trying to gain the lead in pick-up trucks and vans. Already ahead in production over Tesla's CyberTruck, their R1T pick-up and R1S SUV are more traditionally styled and beautifully practical. Rivian have raised over $4 billion. This is a serious electric producer who also plans to build electric delivery vans. Amazon has placed an order for 100,000 electric Rivian vans. Starting its life in the autonomous/ electric car technology space they only began building cars in 2017. Going from just 100 people in 2016 to over 1,000 in 2019, Rivian is on a Tesla-like exponential growth trajectory, they are one to watch.

Next of note is Lucid Motors, with its 'Lucid Air' competitor to the Tesla's Model S; they aim to become a leader in the luxury electric segment. The Lucid Air has similar lines to the Model S and promises a host of innovations. The similarities with Tesla are not a surprise considering that CEO and CTO, Peter Rawlinson came from Tesla where he was chief engineer on the Model S. Many of Lucid's 2,000 employees are former Tesla employees. This is how Tesla's reach and influence have gone beyond just cars, by being an incubator for up-and-coming talent in the new electric arena. Not too dissimilar to the likes of Google, where many past employees have now founded their own companies, Tesla is helping to spread and seed knowledge beyond its company borders. Originally starting with batteries and electric drive trains, Lucid announced, in late 2016, its intention to build an all-electric high-performance luxury sedan. In another sign that big oil knows its days are numbered, the Public Investment Fund of Saudi Arabia invested in the planned expansion of Lucid. Going beyond just a flagship luxury car, Lucid has also announced its intention to build SUVs and a competitor to Tesla's Model 3 by 2025. The big battery Lucid Air is projected to have a range of over 400 miles, 650 km, on a single charge using the same 2170 lithium-ion battery cells as Tesla. Lucid's battery packs have also been used in Formula E (the electric cousin of Formula One). Lucid has the closest DNA match with Tesla and they aim to reach 400,000 cars per year from their new Arizona factory.

Named after the discoverer of electromagnetic induction in 1831, Michael Faraday, Faraday Future is another California based electric start-up aiming big. Building its $1 billion manufacturing facility not too far from Tesla's Gigafactory in Nevada, Faraday aims to deliver 400,000 luxury electric cars by 2026. Current CEO, of BMW i8 heritage, Carsten Breitfeld, joined Faraday to bring some order to past chaos that saw the automaker get into financial difficulties. Merging with Property Solutions in the US and aiming to go public via the controversial Special Purpose Acquisition Company (SPAC) route with a $3.4bn deal, their first product – the FF91 SUV comes on stream in 2022 as another luxury consumer offering.

Without detailing every single start-up, many are adopting the same strategy as Tesla, namely start at the top with a luxury high-end car. The primary reason why luxury vehicles are coming on the market first is the cost of battery packs. Until battery pack affordability allows for cheaper models, the electric revolution will be a class divide. Even the ultimate supercar maker, McLaren, has outlined a plan to build electric supercars. The wealthier members of society will be electric while the poor will be still using petrol and diesel. These start-ups, in replicating Tesla, will help, in part, move the world to zero-emissions mobility.

The issue is, however, that the luxury segment is tiny and represents just a fraction of the billion-plus cars on the road today. The good news is with massive advancements in battery technology the mass affordable electric car is tantalisingly close. Whether all these new car makers will survive and be successful is yet unknown. Many, like Fisker Motors, nearly died already, only to be given second chance via last-minute investment deals to restore their electric dreams. As with every new technology causing a paradigm shift, there are winners and losers, and the best technology does not always win. Having a great car is only as good as your ability to mass-produce and market it. Big challenges, from the global supply chain, battery metal mining, workforce skills and supply delays need to be overcome. A 2021 example was the global silicon chip shortage that slowed production lines. Unexpected problems, like Covid-19 and lack of chips, could derail some of these start-ups given their monthly cash burns, which are enormous and unlike Tesla today are still far from producing profits.

Other established, non-automotive corporations, whose coffers are flush with cash, also see the opportunity in the electric revolution. With existing legacy automakers now racing to go electric, combined with this deluge of new electric start-ups, others are circling. Several non-auto companies are coming in from outside the industry. There is now no doubt the electric revolution is here to stay and can only accelerate. Global brand names like Apple and, briefly, Dyson are also sniffing around the electric feast.

APPLE TO DYSON

Two of the most successful global companies in their fields have both been drawn to the electric revolution. Although their industries and products are far removed from the automotive world, both Apple and Dyson launched electric car programmes. Sir James Dyson, the famous British engineer responsible for the well-known vacuum cleaners, amongst other consumer electric goods, started an electric car project. The UK's richest man backed an all-electric car project with £2.5m of his wealth. With a very successful Midas touch on his other electric products, the expectation was that he could do it again with an EV. In October 2019, however, Dyson cancelled the project saying, "We simply can no longer see a way to make the project commercially viable". A working prototype SUV Dyson was constructed under much secrecy. Dyson became a victim of the timing between EV viability and battery pack affordability. He had proposed a huge 150kWh battery capable of a 600mile range.

Dyson had been mulling the idea of building an electric car since 2014, having witnessed the success of the Tesla Model S. Rightly predicting a rapid future electric transition he felt with his company's background in electric motors, batteries, heating/cooling, and air management they could produce a competitive electric car. The problem for Dyson, and others considering doing something outside their core business, is focus. Tesla's Model S was more of a proving strategy than a profit-making strategy. Knowing it would be difficult to break even with a luxury EV, Tesla used the consumer success of the Model S to attract funding to produce the more profit-making Model 3 and so on. Dyson did not have this luxury and instead would have been subsidising their likely loss-making luxury electric SUV project. The mothership could only sustain such a massive undertaking for so long. Eventually, lower battery costs will make even the more expensive car models 'profit-capable', however, this would not have been fast enough to make the Dyson car viable. Hiring ex Aston Martin chief engineer, Ian Minards, the Dyson project had serious intent. Dyson converted large former RAF hangers to research and development facilities for its EV. With extra-large wheels (for a smoother ride, quiet running, and low rolling resistance) and a strikingly futuristic design, it would have cost £150,000 (over $200,000). Forecasting a low 5,000 sales in its initial year and a transfer of production to Singapore in year two, the project just couldn't make financial sense. Had his 'science project' been fully realised, Dyson would have produced an innovative machine, worthy of his name, but his timing was a few years too early and it's a big leap from vacuum cleaners and hairdryers to road cars.

Smartphones to cars is an even bigger leap. Apple's Project Titan has been the stuff of myth and legend on and off since 2014. At several points during Tesla's growing struggles, Apple was often touted as a potential suitor to acquire Tesla. Great design and innovative leading-edge technology are the similarities between Apple and Tesla.

Unfortunately, that's where the similarities end, as there are virtually no real synergies between the two companies. Apple has built its Apple CarPlay software interface for car 'infotainment' and navigation, however, it is nowhere near a car maker. Tesla has heavily invested in its autonomous car technology and claims to be close to delivering full self-driving autonomous vehicles. Under Tim Cook's leadership, Apple likely recognises the potential for most mobility to be autonomous in the future. With previous Apple divergences into smartwatch wearables, speakers, and health, they can see the massive potential as mobility goes electric and autonomous. Not to be left behind, Apple has reportedly already invested billions in its Project Titan, the codename for an electric and autonomous Apple car. Titan has been veiled in secrecy; Apple hasn't even publicly acknowledged its existence.

In January 2021 rumours circulated that Apple was in advanced discussions with Hyundai / Kia to build an Apple car. Rumours of a $3.6 billion investment by Apple into Kia for the Apple car was wildly speculated. With both sides denying the rumours, industry insiders and automotive pundits worldwide confirmed these talks were taking place. It would make sense for Apple to outsource the construction of its Apple car to an experienced builder like Hyundai, as Apple don't build their iPhones or consumer products, they just design them. The 2014 rumours began circulating after Apple's Tim Cook visited BMW's headquarters to take a tour of the BMW i3 electric car factory. Rumours that the talks with Hyundai/Kia did not come to any conclusion followed stories that Apple had also approached Nissan. Like with new upcoming iPhone releases, the global rumour mill has gone wild with speculation on the potential features of an Apple car. While some of these range from the obvious to the absurd, they all settle on one thing that the Apple car will be a self-driving personal autonomous vehicle. The Apple brand will no doubt bring followers to anything Apple produces at the right price range, however, this project is very outside their core business. As with Dyson, Apple is venturing into unknown territory. Smartwatches and speakers are not a pivot on their business model, unlike building an electric car even if outsourced.

Apple does however have the cash to burn and could be able to sustain a 'science project' for a considerably longer time than Dyson could afford. With possibly a cheaper battery, offering a longer range, it is claimed Apple will become a leader in the electric vehicle market as it is with smartphones. It is hard to see, even with the most extreme confidentiality protocols, that the hundreds or maybe thousands of Apple employees working on Project Titan would not have leaked any revolutionary new battery technology. Elon Musk has said it was easier to launch a rocket into space than building an electric car company. This would be new territory, even if outsourced, for Apple. Steve Jobs was reportedly looking into the car business back in 2008, however, stalled the project due to the global financial crash and bankruptcies of companies like General Motors.

Since 2016 Apple has been investing in companies and start-ups in the mobility and autonomous vehicle space, lending more strength to the rumours. Former Tesla employees, from engineering to software and autonomous divisions, have reportedly moved to work on Titan. Elon Musk even called it "the Tesla graveyard" adding "if you don't make it at Tesla, you go work at Apple". If Apple manages to pull it off and produce a revolutionary design with ground-breaking new battery technology all wrapped in intuitive Apple-like software, they could become the next Tesla and more. However, any new car company will face the same challenges of the supply chain, battery components and access to talent. Like intelligent life in the universe, Tesla could be a one-off in disrupting the slow-moving legacy auto industry. It's been nearly 50 years since the last successful auto start-up became a mass producer. Tesla broke the mould, but it still took 17 years to get to profitability.

Google's Waymo autonomous project has logged millions of miles in development already. Apple has a long way to go before it can get close to claiming a safe, reliable, and fully autonomous vehicle. Assuming all the production and scale challenges are met, Apple is not known for cheap consumer products. It's a premier brand and its products are priced at the mid and upper levels of the market. How much would an Apple car cost? Before answering that Cook should spend some time with Dyson to understand why its $150,000 car simply wouldn't have worked. If Apple used even half of its $190 billion cash hoard (as of 2021) on project Titan, it would dwarf the investments in Tesla and others going electric. They could even buy many of the American automakers if they wished to own production facilities. The iPhone was a revolutionary new way of communicating and computing in your hand. If there is the scope for Apple to make a similarly revolutionary new mobility product, with the money it has, and if they truly have serious intent then they could produce the iPhone of cars.

SOLAR

Electric cars are already far more efficient at using energy to create movement than their energy squandering internal combustion cousins. Ensuring this energy is produced from clean sources is another matter. The generation of electricity still comes in most cases from non-renewable and fossil fuel sources, I'll cover this later in the book. What's the point of zero-emissions cars if fossil fuel electricity stations provided the power? It's still better than internal combustion and later in the book, I will explain how and why it's getting better. However, what if the car could generate its renewable energy from the obvious candidate...the sun? Those crazy-looking one-man solar cars from the Australian World Solar Challenge races you may have seen are of course impractical, but what if a practical everyday solar car could be created? Wouldn't that solve our CO_2 problem from cars overnight? From Elon Musk's perspective, at least, no. When asked about a solar roof for the then-upcoming Model 3, Musk said they were "Not that helpful, because the actual surface of the car is not that much, and cars are often inside. The least efficient place to put solar is on the car." The top of a car has around 3-5 square meters of space, solar is most efficient when angled to the sun, the sun needs to shine, and the cost of solar installations must be economic and look good. With all these constraints, Musk has not entertained a solar roof future for Tesla cars but has intimated it could be a more practicable possibility for trucks and vans with bigger surface areas. Although solar cell efficiency has greatly improved (and a lot more about this later in the book) over the years the efficiency equation doesn't add up. Or does it? This conundrum hasn't stopped several start-ups designing and building electric solar car prototypes with big production plans.

As I was writing this section of the book, the news was released about one such start-up, Lightyear, successfully raising another big investment round of $48m for the development of its luxury segment solar car. I came across Lightyear several years ago. The Dutch start-up has a roadworthy working prototype electric sedan with its entire roof and rear coated in solar cells. To be clear, these start-ups are not producing solar-only cars. They are battery electric vehicles chargeable through the normal methods. The solar bit is more of a free 'top-up'. For Lightyear's Lightyear One, they claim their solar roof can add 30-40 miles (50-65kms) of range per day during direct sunlight in the summer months. The Lightyear One is a sleek, large executive class sedan. Foregoing the rear window in favour of the continuation of solar panels, Lightyear One will have cameras only for reversing and parking. The concept for the car originated from the three-time winners of the World Solar Challenge team from Eindhoven, NL. The story is about efficiency. Aluminium and carbon fibre construction provides the Lightyear One with a super-lightweight body. Advancements in lighter weight materials and extreme focus on ergonomics, to maximise the efficiency of the battery weight to car ratio has been paramount in its design. For a solar car it's attractive, a sleek design with 5 sq. m of solar panels installed on the roof, hood, and rear window. Containing a 60-kW battery pack, capable of delivering at least 400 km, constantly topping up with solar panels. With enough sun the solar array could give this car about 10,000 free miles per year of the range. The car also continues to receive electrical charges from the sun while driving. Lightyear developed their motors, and instead of one large motor in the rear or the front of the car, each wheel has its own motor. This configuration is claimed to be more efficient by Lightyear. Acceleration is not a priority, although it can do 0 to 60 in around eight or nine seconds. Whatever becomes of the Lightyear, it has focused on ultra-efficiency coming from its racing heritage.

The benefit of this technology is likely to be adopted, in some part, by other automakers in the future. With continual improvement in solar cell efficiency, solar will likely feature in some way on your future electric car. When the cost economics work, it is very exciting to think that while parked on the side of the street, on a sunny day, you're gaining free miles with absolute zero emissions. For those living in sunny locations, it is conceivable that Lightyear owners would rarely ever charge their car, except on long journeys. This is not a car for the mass market, with an estimated price tag of around €150,000, it's in the Dyson sphere. This will appeal, as the Tesla Roadster did in 2008, to wealthy buyers, and novelty enthusiasts and may help spark a wider mini-solar revolution.

Aiming at the mass market and car-sharing economy, German start-up Sono Motors is producing an entirely different electric solar car concept. I have known Sono motors for several years. With Sono, the entire car has solar panels. The doors, side body panels, roof and trunk are covered in solar cells. This is not an executive styled sedan, it's a boxy people carrier, a small mini-van concept. The team have, like Lightyear, really doubled down on efficiency. Using lightweight materials and weight-saving features wherever they can to produce a highly efficient machine. Sono claims almost all average daily commutes could be powered by the sun, even in winter. Such is the public interest in a clean machine like this, that when Sono motors launched a crowdfunding campaign, it was extremely successful raising over €50 million. Sono believes the future is solar and shared, the car has been designed to be a perfect city candidate for short-hop car sharing. With an innovative car to grid capability, and even a 220v domestic plug outlet onboard, a Sono car could part power your house.

Named the Sion with a target price tag of $25,000 or less, this car will be mass-market capable. Thanks to a bi-directional charger the Sion will be able to power other electric cars – in theory, a Solar Sion could fully charge any EV given enough time and sunlight. Both Lightyear and Sono are pioneering companies, at the leading edge of solar car technology. It may be a while before they become mainstream and practical at the right price point. Other pure electric car start-ups are also beginning to incorporate some solar features. Fisker motors planned Ocean, which the company claims are "the world's most sustainable vehicle", has a solar roof.

The Fisker Ocean's roof can generate a "1,000 free, clean miles per year" according to the company. The SUV is expected to be priced under $40,000 per year and come on sale in 2022.US start-up Aptera has also announced an electric solar-powered car. With a far more radical design, these three-wheeled space ages looking cars are claimed to never need charging. The company did launch an innovative calculator on its website, which enabled people to set their location on the world and estimate how much sunlight they will receive and hence free range. Another solar start-up was created by former Lightyear employees who formed Squad – a golf-cart sized solar city car maker due to come on the market with a price tag of just $6,000.

Appearing to eventually acknowledge the potential of solar, even if limited, Elon Musk has offered a solar roof option on the upcoming CyberTruck providing around 15 miles of range per day from sunlight. Not limited to just the surface area of the vehicle, fold-out solar 'wings' have also been touted as a potential solar capture solution. Many concepts have been developed to act like convertibles, with the solar 'wings' folding out of the trunk, in a satellite dish fashion. These start-ups are not the first to take a serious look at solar. Toyota, which today are way behind in the EV race, did incorporate a solar roof in its Hybrid Prius. Only usable while the car was parked, and adding just 3 to 4 km per day, the feature was more of a top-up novelty. Newer versions of the Prius, with more efficient solar panels, are claimed to deliver up to 30 to 40 km of range per day.

Hyundai also offers a solar roof on some of its models as an option. I will cover later the challenges of EV charging at home, particularly if you don't have access to a charge point. For average daily commutes, solar charging could be a practical solution. Solar cells do add weight, cost and complexity to a car, however, this is offset by the absolute zero-emissions gain. It looks like solar, in some form, is here to stay. As solar cell technology and battery efficiency continue to improve, a percentage of miles driven by sunlight is likely to become a normal part of our future driving.

BUSES, VANS & TRUCKS

The CO_2 problem isn't just from our cars. The movement of goods around the world by ship, plane, truck, bus and van all contribute to our climate crisis. While many countries electrified their rail networks decades ago, buses, in most cities, are usually heavy polluting diesel machines. Indeed, the most polluted city street in the world at one point was London's Oxford Street. This was in a large part due to the hundreds of red double-decker buses traversing this busy thoroughfare daily.

Over a decade ago I lived in London for several years. I hated the confinement of the Tube, especially on hot evenings making my way home from the city. It was just under one hour walk from my office to apartment and regularly I would walk home. This was long before any electric cars were on the streets of London, and before congestion charges were introduced. Walking home, at rush-hour, the city was jam-packed with cars, buses and vans. I do remember on a few occasions coughing heavily when I got home and the 'nose soot'! I was oblivious back then to the health effects and climate impact of this extreme air pollution. If I had had my air quality meter back then, I have no doubt it would have been off the scale. The hundreds of black cabs and buses were, and still are, mostly responsible for this. London is just one example, but I've been in many crowded European and Asian cities, where you are near choked with the smell of diesel fumes. Diesel engines are the ideal heavy workhorses for the transport of people and goods, however, they must go and quickly. With cities introducing tighter emission standards and future ICE car bans, many companies in the goods transport sector have realised the writing is on the wall.

In addition to being forced to do this, being seen to be clean and green is now an essential part of any corporation's mission. Companies like Amazon and UPS have made commitments to radically reduce their carbon footprints. To date, however, there's been very little progress in the development of electric freight in comparison to passenger cars. This is about to change and rapidly. The last-mile delivery market is witnessing a late but accelerating electric revolution. Millions of heavy diesel buses around the world transport people at the expense of air quality and climate; this is all about to go.

Several start-ups as well as existing manufacturers are now about to release a slew of electric buses. Arrival is just one of those start-ups. Headquartered in London, Arrival has been developing electric bus and van concepts for several years. Back in the early 1900s, London had a fleet of electric delivery trucks – only to be replaced with diesel by the 1920s. Arrival's goal has been to produce vehicles that are as cost-competitive as their gasoline equivalents. In October 2020 Arrival raised a further $118 million from US investment firm Blackrock, this is on top of their Hyundai investment of over $100 million. United Parcel Service placed an order for 10,000 of Arrival's future small vans and options for another 10,000. Freight electric vehicles deliver less range than passenger cars due to the weight to load ratio, of around 200 km. These vans are squarely aimed at the city market where deliveries take place during the day, less than 200 km of driving, and charging takes place at the depot at night. Ironically, given they were wiped out with the arrival of the iPhone, Blackberry technology will power part of the Arrival vans. Now a British unicorn (a company start-up valued at more than $1 billion) with a valuation of close to $4 billion, Arrival is on a path to being a huge UK electric success story.

Conversely to Tesla's Gigafactory model, Arrival is building what they call micro-factories. These factories are aiming to be ultra-efficient with low-carbon footprints producing components in multiple different locations, for a single assembly at one location. This is a production method used by the giant aircraft manufacturer Airbus. This is an interesting strategy and perhaps more suited to the production of vans given they will not be built in the millions of units as with passenger cars. Offering a paradigm shift for delivery companies, Arrival has added a range of features to help the driver make their delivery easier. Just as with all other EV start-ups, they plan a future where their vehicles are autonomous. Converting all commercial vehicles to an electric fleet in big cities will have a dramatic impact on noise and air quality. Even without new start-ups, the likes of Mercedes already offer eSprinter and eVito van options, built on the old ICE platform but now with shoehorned batteries and motors. On top of the environmental benefits, for the logistics companies, the maintenance costs are lower. Typical delivery van fleets get a lot of abuse on a day-to-day basis around city streets with constant start-stop driving. Apart from tires, much of the normal wear and tear of an internal combustion van or bus is gone with electric. I can't wait to walk down a busy London or other major city street and have near silence and clean air. Arrival's electric bus is also a striking futuristic public transport vehicle. With side walls covered in LED panels for outside advertising and large, almost floor-to-ceiling windows, Arrival's bus is striking and will be a beautiful fresh addition to city streets.

Once again this is not the first new electric bus company, this story goes way beyond Arrival. Thomas Edison invented an electric bus back in 1915. Borrowing the Edison name, a Korean electric bus start-up has also developed and is now producing electric buses. In China, BYD, have produced an electric bus and many major Chinese cities are making massive strides towards full electrification of their public transport infrastructure (more on China later). Others, like US-based Proterra, are also developing electric buses. Given the smaller volumes, there won't be anywhere near the same number of new start-ups in the bus market as with passenger vehicles, however, the choice and range are increasing, and you can expect to see your city gradually move electric. Intraday charging for the heavily used EV buses has also produced some innovative solutions. For example, in Luxembourg, overhead power connectors have been installed that attach onto the roof charging 'port' while buses are parked at a terminus.

On the heavy truck side, it's not just Tesla's Semi making waves. There has been a host of new start-ups in the race to electrify the semi-truck. Existing truck builders have also launched 'ST' development programmes. Daimler's Freightliner eCascadia electric truck is one example. You may think the batteries need to be enormous for long-distance road freight transport, however surprisingly in the US, 80 percent of freight is transported less than 250miles.

Electric trucks are viable with predictable routes and very definable charging at rest stops along major highways. A bigger challenge is the electrification of freight depots and logistic distribution centres where loading and unloading can be done in less than 20 minutes. China's BYD, with an assembly factory in Lancaster, California is rolling out electric delivery (mid-sized Class 8-day cab – i.e., larger vans) trucks in the US. Another Chinese-backed start-up, called Chanje, has developed a medium-duty van with a 3-ton payload capacity and 150 miles of range with over 900 on order for FedEx already. Troubled start-up Nikola (so named after Nikola Tesla), whose CEO resigned in 2020 amidst claims of exaggerating its hydrogen technology, has created the Nikola One, Two and Three for the US and Chinese Markets in a full battery-electric or hydrogen fuel cell configuration (more on hydrogen technology later). One of the mainstays of trucks, Volvo, has also developed a zero-emissions offering called the VNR. Already on the pilot on Californian roads in partnership with California's South Coast Air Quality Management District (SCAQMD) the company strategy is to prove, then scale its technology. Electric van start-up, Workhorse, has also come on the EV logistics scene with its C1,000 and 950 orders from UPS.

President Biden's January 2021 pledge to replace 650,000 government vehicles with electric further accelerates commercial electrification. Whilst somewhat behind the passenger car market in terms of development, trucks, vans and buses are making a rapid sprint to the finish. With 25% of CO_2 emissions coming from moving goods, this is a very positive development.

CHINA

China has grown to become the world's second-largest economy and largest CO_2 emitter in just 30 years. Astonishing economic growth has seen the country of 1.4 billion people become a global economic superpower. With rising wealth and growing consumer demand for a 'western lifestyle' China will overwhelm any climate CO_2 targets without radical action. We have witnessed large scale public infrastructure development at lightning pace with Chinese cities virtually springing up overnight. Roads, rail, airport, and military progress have been staggering. The ability of the Chinese to mobilise en masse and achieve tremendous progress, in a relatively short space of time, could bode well for tackling their CO_2 problem. China is doubling down on electric. The electric developments in China are also acting as a catalyst for other countries, tipping the balance towards our electric future. Apart from the many Chinese EV Start-ups, state-wide electrification programmes are underway in a massive, concerted effort to be a world leader in electric mobility.

Domestic electric car makers are taking some market segments by storm. Indeed, the Tesla Model 3 was outsold by a $4,600 Chinese made mini electric car in 2020. The Hong Guang mini EV has a range of just 120kms and only has a top speed of 100km/h however it has proved very popular in congested Chinese cities, like the Mahindra mini EV in India. To incentivise electric car take-up, the Chinese government offers free license plates for EVs, this contrasts with trying to get a license plate for a gas-powered car which can be problematic as the government carefully manages the license plate 'lottery' to reduce city congestion. The mini EV was the second best-selling electric car in the world in 2020. Going way beyond just tiny city cars, the Chinese government has made a massive investment in charging infrastructure. Now the world's largest market for electric cars, the government is also offering big subsidies for electric over ICE.

The world's largest emitter of greenhouses gases has woken up to the challenge of converting the 400 million ICE cars on China's roads to electric. China, despite its large size, has few natural resources and is heavily dependent on oil imports. Reducing foreign dependence for the future global superpower is a key policy of the Beijing government. With just over a 1% reduction in oil imports to date (thanks to electric car adoption) they have a long way to go. Beyond just passenger cars, a big push is on to electrify public transport. Chinese company BYD has already completed an electrified bus network in Shenzhen. A city of over 12 million people, Shenzhen now has over 16,000 electric buses and 22,000 electric taxis on its roads covering more than one million clean kilometres per day. This is the future template for all Chinese cities, 100% zero-emissions public transport.

Supported by almost 2,000 chargers their bus network made the transition from diesel to electric and it works, moving over one million people per day. The transition to electric, with charging downtime, did not impact the daily utilization rates for the buses, indeed it improved with lower maintenance, hitting over 99% uptime on average. The taxi fleet is comprised of BYD's e6 cars. Purpose-built charging facilities have been built, where taxi drivers enjoy recreation and rest complete with gyms, massage facilities and food courts while their taxis charge. With some of the worst historical air pollution in the world, Chinese cities can now demonstrate a world leadership model in making quick moves to workable electric mobility.

This is globally achievable and requires the recipe of government incentives, rapid deployment of charging infrastructure, EV start-up support and disincentives to drive internal combustion vehicles. Unlike Europe and the US, China doesn't have the hangover of big automotive industry unions or legacy ICE manufacturing practices, as it never really had a historical automotive industry. Old school European and US manufacturers, while they built some very high-quality vehicles in the past, are contaminated by the legacy of their past. This means, as, with any major legacy business, it's hard to get truly 'out of the box thinking' like a start-up mentality, like Tesla and NIO have. China has none of this legacy, allowing it to move fast and deliver innovations rapidly. New Chinese EV companies like NIO, which is in the luxury segment, are offering their customers unique and innovative services such as drive-thru battery swaps and club memberships. Their cars even boast an internal AI assistant that looks like a little robot on the dashboard.

China also doesn't have large domestic dealership networks where individual dealer profit and loss is king. In many cases, these legacy dealerships are relatively independent of the incumbent manufacturers. This means they sell the cars that sell the best and easiest, and this could act as a disincentive to promoting new electric cars with smaller margins. Without any legacy dealership bias, it allows the promotion of electric cars to consumers to be prioritised. There is even further resistance to change when it comes to car maintenance. Dealerships and legacy automakers have lots of relationships with car repair services and maintenance third parties. With an EV the maintenance is largely just tires and brakes, with none of the normal components needing replacement or servicing. On top of this with most EVs fully connected and always on, diagnostics highlighting faults and repair scheduling is almost completely independent of the driver. This means there are no protected relationships in the consumer chain as there are with internal combustion. China is therefore uniquely positioned to go all-in on electric and forget about legacy.

China wants to be the world leader and has audacious plans to accelerate its EV transition. In a 2021 study by McKinsey research, a poll found that only between 10% and 30% of Americans think their next car will be electric, that number is 40 to 60% in Europe and over 70% in China. Sales of electric vehicles in China grew by over 60% in 2020 to 1.2 million. There are also close to 1,000,000 chargers now in operation, of which over 300,000 are public charging stations. This is double the number in the United States. On top of customer subsidies, the Chinese government also set quotas for electric car manufacturing and encouraged overseas players to enter the Chinese market, like Tesla and its Shanghai Gigafactory.

The United States, as the world's number one superpower, would like to hold on to that title. The US is currently 2nd to China in the race to fully electric mobility. With the Biden administration, there is a renewed focus on climate issues and the US aims to catch up with China. China does not intend to stick with just its internal market and, as anticipated, they will aggressively enter overseas markets. The electric car race has the potential to influence geopolitics. China's race could give them a global dominance in the electric revolution. China is currently leading in the development of lithium-ion battery mega factories. In 2022 of the more than 140 new battery plants under construction globally, China accounts for 107 of them. This compares to just nine in the US.

It is estimated that automakers will invest over $300 billion over the next five years in EV development; more than half of this amount is in China alone. In a testament to the belief in China's EV ambitions, legendary investor Warren Buffett heavily invested in Chinese BYD. With an 8.2% stake, Berkshire Hathaway owns more of the Chinese electric automaker than it does of General Motors in the US. China hasn't enjoyed great success in the electric car export market yet. China only exported 1% of its electric vehicle production in 2020, compared to Europe's 17% and the US's 30%. This is about to change, however, with Nio, Xpeng Motors and BYD beginning to export their cars into the European market. If this doesn't further incentivise the European players to accelerate their EV plans, then they could get left behind. China has set a carbon neutrality goal by 2060 and it's on track to deliver a 20% decline in CO_2 emissions by 2035. The Chinese EV revolution is of course only part of the carbon dioxide solution. China is also looking to transition much of its energy away from coal and oil-powered power plants into renewable sources; I'll cover this green energy transition later in the book.

Tighter emissions standards and Covid stimulus policies in Europe have allowed EVs to gain an 11% market penetration in 2020, this beat China where just a 5% penetration was achieved. Although Europe has a market share lead when it came to electric vehicles, the trajectory of growth for China is far stronger. When it comes to global sales of electric cars, of the top 13 manufacturers, four are already Chinese brands, with BYD second only to Tesla. When it comes to the batteries inside the world's electric cars, batteries from Chinese suppliers represented just 10% in 2012 growing to 50% by 2020. According to the International Council on Clean Transport, ICCT. China's battery production capacity is likely to keep it a global leader to at least 2025. In further evidence of China's lead, 40% of the world's electric car sales were in 25 global cities, 14 in China alone. With the rest of the world catching on, it's uncertain if China will indeed lead the EV revolution in the future. In 2022 Norway, The Netherlands, UK and California have the highest share of electric vehicles of new car sales, with China coming in 5th place. While China has made tremendous progress, its new EV sales targets are less ambitious than the likes of Norway Canada and the UK. With these countries aiming to be banning the sale of ICE cars 100% from 2025/30 onwards, China has only set a 20% target by 2025.

Right now, license plate auctions, parking space incentives, new car subsidies, automaker quotas, backing new EV start-ups and the added benefit of no real legacy auto industry has given China the global lead in the EV revolution in total numbers currently. Europe and the US can only look on and try to model some of China's policies and actions. There are inconsistencies in the Chinese story, however, whilst they have pledged to reach carbon neutrality by 2060, massive development in oil and gas transmission pipelines is still taking place in parallel. The plan for the world to reach zero carbon emissions is the key goal in battling our climate crisis. China's strong lead in transitioning to electric cars is to be commended, and many cases should be replicated by other countries. However, matching this with clean energy production alternatives may be an even bigger challenge for the Chinese as they vie for a global leadership role.

I'll explore the transition to clean energy combined with the electric revolution in more detail later in the book and who may be the likely winners and losers.

PLANET EV

There is an entire ecosystem of new start-ups popping up across the planet vying to get in on the EV revolution action. As with all historical technology leaps, there will be winners and losers, acquisitions, mergers, breakthrough technologies and dead enders. The EV revolution is not just confined to cars, later in the book I cover some incredible electric advancements in commercial aircraft, trains, micro-mobility and even flying taxis. For now, I'm going to cover how the tipping point to EVs is not just about making cars. Like the space race – multiple new technologies and innovations are being born out of the revolution. If the fickle financial markets are to be believed, then the move of money into EV related investments massively increased in 2020 and 2021. Investment banker Morgan Stanley, in 2020, profiled over 60 companies that have a significant interest in electric car adoption. These names are not just carmakers, they include silicon chip manufactures, mining, chemical companies, battery technology companies, charge point operators and even big oil divisions. Although global EV penetration as of, 2022, still only remains at 23%, Wall Street always goes where the money action is and it is a clear sign that the investment community smells the coming EV action.

There is a fascinating array of new start-ups appearing globally. Certainly, California is a key hub for electric car innovation, however, there are other global centres. Israel has emerged as a global EV hotspot. The small country has spawned several highly disruptive technology start-ups. One such start-up attained great media attention in early 2021 with its announcement of revolutionary new battery technology. Store dot has claimed to have developed a new generation lithium-ion battery that can be charged within five minutes. I'll cover new battery technologies, including future solid-state batteries, later. Another Israeli start-up is REE, has developed a completely different way to develop the floor plate of electric cars. With its design, everything is integrated into a single universal floor plate, with independent motors at the corners. Each 'corner' of the floor plate acts independently on each wheel and can be maintained individually. REE claims that this will make manufacturing more efficient while allowing for a greater and easier variety of car models to sit on top of its platform. Other Israeli start-ups such as Driivz, providing charging management platforms, ElectReon, for wireless EV charging and Addionics, for 3D battery pack design, are blazing an EV innovation trail.

In the UK, battery maker Britishvolt is developing a $3.5 billion battery production facility in northern England while in the US Sila Nanotechnologies is specialising in increasing the volumetric energy density of batteries, reducing the cost per kilowatt-hour. These are not just fly by night start-ups. Sila has already raised almost $300 million. In Germany, Quantron specialises in the electrification of existing ICE vehicles while in the UK EO Charging has developed the world's smallest smart home charger.

I am not going to name every single start-up in the EV space, but the above gives a flavour for the diversity and scale of what is happening in the background. There is also a host of software start-ups and mobility-as-a-service start-ups in the electric car space. Whilst my own UFODRIVE company was the first to go all digital and all-electric, we are no longer alone. There have been a host of car-sharing platforms launched across the globe in just the past three years. The future of car ownership is changing rapidly, in part because of electric cars.

Worldwide EV momentum is growing, with an entire EV ecosystem now also emerging and an unstoppable race towards our electric future underway. Global sales of electric cars are now 10x the level of just five years ago. President Biden's $174 billion investment in EV development, announced in mid-2021, is testament to the level of focus now on electrification and decarbonisation. Prediction models for electric car adoption from only four and five years ago are already out of date. These models are constantly being revised; investment bank UBS predicted in early 2021 that they now expect 100% electric mobility by 2040. I believe it will be quicker, I predict that close to 100% of all new car sales will be electric by 2030. On top of consumer awareness, there are numerous city bans for ICE coming into force and more and more are shifting to electric and clean mobility.

Just like the smoking bans globally now alienate anyone who smokes indoors – it will be the same for electric. In the future driving a heavy polluting diesel, petrol and later hybrid will be frowned upon and those offenders are likely castigated as smokers are today. These factors, combined with the rapidly improving economics of owning an electric car and the flood of new models coming on to the market will further accelerate adoption.

Who will be the winners & losers? ICE resale values will plummet from the mid-2020s. They will become legacy technology. This will be driven by better ownership economics, better customer experience, regulation, and climate consciousness. Legacy automakers were oblivious and indifferent to Tesla's rise, leaving them to a virtually uncontested electric market. Globally, titans like VW and GM are now desperately racing to catch up. The only survival route for all legacy automakers now is to create independent electric operations with start-again thinking. Push all brain power and fresh ideas into these divisions and then wind down the internal combustion operations over time. Any Frankenstein effort to combine the two businesses will hinder progress with hierarchical bureaucracy and management layer self-interests. If they go down this route Tesla, who have already eaten their lunch, will eat their dinner too.

There are internal civil wars in the legacy automakers, as the battle for good and evil wages between electric and internal combustion. The 2010s was Tesla's decade. Tesla changed the public perception of electric cars, and with further education, splitting facts from fiction, more and more people will begin to buy electric. Once you go electric, you don't go back. A 2020 survey shows less than 1% of EV owners want to go back to gas according to Electrek.com. In the 21st century, a company that fails to adapt and change fails to exist. Electric car sales were up 40% in 2020, at the same time global car sales dropped 14%. US carmakers will spend over $250 billion chasing Tesla over the next three years.

By 2030, Tesla will be just another automaker, everyone will be electric. Will they be the Toyota of the automotive world with the greatest numbers of car production per year? – it's very possible given their already giant head start. Others, like Volkswagen, will catch up and may even overtake. Regardless of which companies make it and which don't, we will all be the beneficiaries. Clean air, quieter streets, and a big contribution in reducing carbon dioxide emissions to save our planet will be the reward. The tipping point has happened, the world has shifted, and the next generation will be fully electric. However, there are still over 1 billion internal combustion cars on the roads, and consumers must embrace electric for these to disappear. With over 90% of cars sold every day still powered by fossil fuels, there is still a long way to go before everyone thinks electric. All the ingredients are now in place, and soon electric cars will be mass affordable.

For everybody to buy an electric car, and only think of an electric car, all the misconceptions, facts and fiction need to be addressed. Electric cars work today, even for long journeys. Batteries will improve, range and charging will get better, however, right now the public still needs convincing. In the next chapter, I'm going to clarify all the facts and fiction around electric cars. Later I'm going to go into some detail on battery, range and charging and how we can transition our whole world to clean mobility powered by clean energy.

CHAPTER 6

FACT & FICTION

FUTURE FEAR

I give ICE cars 5 years to live (2027) in wealthier countries and 10 years in less wealthy nations (2032). This is my estimate for new car sales, not the total replacement of the world's fleet. The average life span of a car is 13 years, so there will be still gas cars in everyday use in 2040, although I expect by then they will be dwindling to a tiny fraction. A big part of the reason I wrote this book was that I was constantly astonished by the strongly held misconceptions about electric cars. The more I got into the electric world and the more people I met in the mobility industry, the more I realised how little people knew, which is ok, but worse, how they had come to believe some outrageous EV mistruths. Time and time again I would hear definitive statements made about all the negative limitations of electric cars from people that had never even driven one. Even worse, sometimes I would be in meetings with very senior big oil and automotive executives who discredited electric mobility and dismissed what I was saying out of hand.

I have many stories. I realised that most people today still have no real idea about how electric cars work, how they are made, what's good or bad and what's fact or fiction. Before I get into these in detail an important point to note. I am not a blind cult-like evangelist that can see no wrong with electric cars, bent on persuading you to my point of view. That's not me. I'm going to instead tell it exactly as I see it, having experienced, read and learned from thousands of first-time electric customers (with UFODRIVE) over five years.

EVs are not perfect, charging networks are not perfect, batteries are not perfect, and I will outline why. However, I will also outline why overall and in balance, EVs now represent a far better and practical alternative to internal combustion for you, our air and our planet. Another point to note is that I don't really care about cars. I like them a lot, however, I am not a 'petrol' or 'electric' head. I love driving my Tesla, and it works 90% of the time for me, but if something better came along, I would ditch the Musk machine. I have had some jaw-dropping statements made to me about electric cars, from people who know nothing about them, with absolute conviction that they are right when they are usually wrong in most cases. Many times, it's been funny and ironic and mostly good friendly banter, but also sometimes a fierce argument. I'd like to tell a few stories from friendly to adversarial before I deal with the facts and fiction. Let's start with the friendly banter first, names changed for anonymity....

To get funding for our electric mobility start-up we first sought money from 'business angels' we knew. Angels are generally individuals willing to invest at the earliest, often pre-launch, stage of a company, in turn for a stake. One such angel, let's call him John, is a friend/business friend of mine I have known for almost ten years. We meet once every few months for dinner or beer. I pitched the idea for our electric rental company to him and he loved it. A few weeks after, he called me up and committed to investing in our start-up, even before we had launched. The size of his investment was not insignificant for him, but he believed in the idea.

Between one thing and another, I didn't see John for nearly a year after that until we met for dinner. Over dinner mentioned he had decided to change his diesel car and had “finally decided to take the plunge”. I said what “do you mean”, his answer was “I have decided to buy a hybrid”. John is a 50-year-old smart guy, high up in the global bank he works in and ‘world aware’. I replied, “why don’t you just buy a fully electric car” (I explain later in this chapter my view on hybrids). John laughed out loud and responded to me in a friendly but condescending tone “ah come-on, sure I drive up and down to Brussels from Luxembourg regularly and I would never take the chance with an electric car!” I was floored by his automatic response and proceeded to highlight the irony of his statement, “John, you do realise you invested a decent chunk of your savings into a fully electric mobility start-up, don’t you?”, his reply “I know, I know but I wouldn’t be ready for that myself yet”.

The example journey he chose, Luxembourg to Brussels is about 220kms (136 miles). This is well within the range of practically every electric car on the market today. What’s even better is it is well serviced with fast chargers along the route and at each end if you need to make a same-day return trip. I have driven this route hundreds of times, in an electric car, with zero hassle. If I could pick one of the best routes for regular electric usage it would be this route, the E411 highway between Luxembourg and Brussels. John is not stupid, he’s not ignorant and hasn’t a malicious bone in his body, he was just unaware, and worse, believed all the myths he had heard about electric cars without question. I set out to change his mind, but I knew I was fighting against years and years of misinformation and untruths about electric cars. John politely listened to all my arguments, nodded his head reassuringly, however, about a month later he texted me to tell me he had bought the hybrid car and that maybe his next one after this would be fully electric. I failed to convince someone who had committed their hard-earned money into an electric car company. If John couldn’t be convinced to go electric yet, how could anyone else? This was a friendly story of course, however, I’ve also been on the receiving end of a far more aggressive and anti-electric audience.

In mid-2020 I found myself on a Zoom call with several very senior investment and information technology executives from one of the world’s biggest oil companies. This meeting had come about because during a pitch to another venture capitalist they mentioned that they had been working with this oil company on their plans to shift some investments into 'clean tech' investments. The online meeting took place across several different countries, including Spain, Great Britain and The United Arab Emirates. The meeting started well, with me and my co-founder outlining all the benefits of our software and our platform to manage an electric fleet, however, then took a suddenly aggressive and hostile tone. I had just commented that the age of the internal combustion car was rapidly coming to an end, that electric vehicles offered a better alternative for drivers and the planet. I hadn't forgotten that I was talking to oil executives, I knew what I was doing and didn't set out to provoke them or seek a negative reaction. However, a negative response I got, in a big way. One of the executives, let's call him Mike, was a 25-year veteran of this oil company. Almost the second I had uttered the words about the virtues of electric cars, he jumped in and made the following, quite extraordinary statement, "With all due respect Aidan, I don't think you grasp or understand the incredible convenience of driving into a petrol station and filling up with petrol or diesel versus having to spend hours and hours charging an electric car". He continued, "Electric cars are simply not practical and will never be mainstream". He then preceded to grudgingly laugh off my belief in the future of electric cars by finishing with this sentence, "We are certain internal combustion cars will represent the majority of cars for a very long time to come."

I couldn't help taking a little bit of personal offence to his tone, he said it in an attacking manner, but I was able to bite my lip, after all, we were the ones looking for money. We never spoke to this company again and in some ways, I'm kind of glad. Mike is one of the reasons this book has been born, his response was a mixture of total ignorance about electric cars, also genuine defensive protectionism of his old, and eventually to be gone, oil industry.

"Petrol cars are grand!". These were the words issued to me when trying to convince a Dublin Airport executive to give our start-up some parking spaces to trial our electric car rental service. I spent much of 2018 and 2019 visiting locations, airports mostly, to try to get approval to open our electric car rental service. Airports are notoriously difficult to deal with, usually public or semi-state run, they are filled with bureaucracy and cumbersome, often protectionist, tender processes. That said, out of the 22 airports I negotiated with, 17 said yes, 4 said yes but not yet, and only one said no. Dublin Airport. I could not convince or navigate the bureaucracy to win Dublin airport. The airport executive even said, "When people arrive, they want reliable gas cars, preferably diesel – no one is going to want to rent electric". To be fair to some of the staff of the Dublin Airport Authority, some of the more junior members of the team did try to help us navigate the civil service minefield, in the end, we had bigger fish to fry and paused the Dublin Airport project. Frustratingly, we did open in Dublin City centre and have had lots of happy electric customers asking when we can open in Dublin Airport.

These three stories represent just a fraction of the hundreds of interactions I have had over the past five years with people from all walks of life. The vast majority simply have no clue about electric cars, what's going on, how quickly we will shift, and what it means for them. They are being bombarded with mixed messages from oil and car companies, and a host of often outlandish claims from electric car makers.

I want to use this chapter to explain, as simply as I can, what is real and what is false. Although I am passionate about electric vehicles, I know there are many limitations, and I still fail to convince many people of the virtues of electric cars. "I wouldn't be able for an electric car" – these are the words my mother, in her mid-70s, uttered to me when she was changing her car in February 2021. She had an eight-year-old diesel car, wanted to change it, and asked me what she should get. She is perfect for electric. Her old car had just 26,000 km on the clock - after eight years! She just drives around the city mostly, never more than 20 km a day. When I told her, she should buy a small electric car, that they were easy to use and she would never have to even go to a gas station again, she baulked at the idea. Fear of the unknown gripped her. I told her I would get a charger installed in her driveway and I even borrowed an electric Hyundai Kona for a day and drove her around in it and let her try it out. She loved it, was amazed by the silence, the comfort and all the convenience electric brought. Two weeks later she bought a Toyota diesel hybrid. When I asked her why, she said she was afraid of electric cars, and it would be easier just to stick with what she knew. I can't of course criticise my beloved mother, she is an incredible woman and open to almost anything new except choosing an electric car in her 70s was just too far a stretch. Most of us are the same, old habits are hard to break, usually, it's a lack of knowledge, fear of the unknown and contamination by myths. This is part of the reason why I wrote this book. Time and time again I met people saying the same old lines about electric cars that they had read or heard somewhere. They usually go something like this:

You can't drive long distances in electric cars. There is nowhere to charge. The battery will die after just a few years and must be replaced. Electric cars are too expensive. It's more expensive than petrol to run. Electric cars are not as clean as petrol and diesel. Slave children are mining the battery minerals in poor countries. They're too complex to use. They can catch fire easily and are dangerous. New models are coming out so quickly that the resale value is terrible. It takes too long to charge. There isn't enough power to charge all these new electric cars. They can short circuit in the wet. They are slow. Technology is changing so much that if you buy an electric car within a year, it will be totally out of date. Hybrid cars are better because they can self-charge and don't need to be plugged in. (I will deal with this false statement later in this chapter). I could go on and on. I've also heard claims made by electric car manufacturers which are sometimes outrageous and fuel the misinformation fire. The range capability, ease of use, even safety ratings are often exaggerated. I'm going to deal with much of this fact and fiction, roughly in the order of which issues I hear most about, starting with electric cars don't have enough range.

RANGE

How many miles or kilometres do you drive on average per day? How many times do you go on a long-distance or multi-day road trip?

When you fill-up your car with gas in most modern cars you will see a range estimate. This indicates an estimated (theoretical) number of miles you will get for your full tank. Practically no one pays attention to this number and the estimate is seldom achieved. Before I bought my first electric car, I was lucky enough to have a big heavy BMW diesel X5, provided to me as a company car at that time. When I filled it to the top with diesel, the range estimator often showed as high as 1,000 kilometres. I never got close to this number, ever. How you drive, how fast, how many passengers, how much luggage, even the temperature affects how much fuel you use.

I drove a 'petrolhead' in my car once; this guy was at one point my boss and collected sports cars, I remember telling him I never got even near 1,000 kilometres out of the tank. He reminded me that the number is kind of irrelevant because the only thing that controls how much fuel efficiency you get is your right foot. This is the same in electric cars. The harder and faster you drive, the more gas you burn, or kilowatts you use. The average range of a gas car is 413 miles/663 kilometres on a full tank. Having driven an electric car thousands of kilometres across Europe and having analysed the data from over 10 million kilometres of UFODRIVE electric rental customers, I am confident in making the following statement.

The range of an electric car is irrelevant... for most use cases.

Range dominates when it comes to EVs in the media and despite my above statement, you will see advertised, constantly, electric cars with more and more range. There is constant talk of the 1,000-kilometre (600 miles) battery for example. This is completely unnecessary. Indeed, the average range of electric cars today, at around 300 kilometres, is plenty for almost everyone's everyday use. Unless you are living in countries with virtually no highway charging network, then this is even enough for most long-distance travel. In the next chapter, I will cover in more detail long-distance charging and range, including the multitude of new fast-charging networks popping up across the globe.

I've also been witness to multiple statements, from all sorts of people, who say that when battery range reaches close to 1,000 kilometres, they will then consider buying an electric car. I would bet that most of them have rarely ever even driven more than 1,000 kilometres (600 miles) in a day in their gas cars. The comparison between the range of electric cars and gas cars, charging or filling up is an apples and oranges comparison. It's the wrong way to compare and add resistance to moving away from gas cars. We must stop comparing the range of a petrol and diesel car to an electric car. The use case is completely different. This is because when we get close to running out of fuel, we go to our nearest gas station, or when we're on a long journey we fill up usually when we still have at least 1/4 of a tank in case we run out. Range anxiety, as it is known in the electric car space, has existed with internal combustion cars since the beginning.

There's a pretty good chance that you have probably run out of fuel at some point in your life. I have a couple of times. I remember one occasion when I was completely distracted on a phone call driving home from work. I knew when I left my office in Dublin city centre that I needed to get petrol on the way home (it was a 40-kilometre drive) I completely forgot. Realising in panic when on a major highway, with no turn off for 15 kilometres, that I probably wasn't going to make it. The range predictor had already gone to zero several miles back, and that horrible feeling of dread began to sink over me. I drove for at least 10 kilometres with the gas tank on empty before the car began to suddenly slow, I just had time to pull into the hard shoulder. I had to call my wife to get a petrol can and bring it down to rescue me.

On another occasion, when I was much younger, and on one of my first road trips, I ran out of petrol in my Volkswagen Golf, as I was rolling into a petrol station in Galway Ireland. Hilariously and my friend had to push the car up to the petrol pump. I know several people who carry around spare 5-litre cans of gas in their trunk, just in case. In almost every single developed country on Earth, unless you're heading out into the wilderness, in most cases there is no excuse to ever run out of gas. It takes just a little bit of planning if you're doing a long journey in a gas car. And it takes just a little more planning if you're doing a long journey in an electric car. A little bit of advanced planning is all that is required with electric.

I gave a talk at the Irish EV Association annual event in 2019. My speech followed a speech by a gas station operator, who was talking about installing electric charging in their gas stations. During his speech, he said the island of Ireland needed thousands of fast chargers, all over the country, before people would adopt electric cars. He was wrong. At UFODRIVE we had analysed the driving behaviour and patterns from thousands and thousands of kilometres driven in our electric cars across Europe and on the island of Ireland. We could see the weak points when it came to range and charging, based on customer experience, and pinpoint where best to place charging networks. When it was my turn to speak, I, in a polite and friendly manner, corrected the previous gentleman by opening my speech with the following... "we don't need thousands of fast chargers in a small island like Ireland, we only need about 20 fast-charging stations roughly in a line down the centre of the country". My opening line got a few gasps and giggles to which I asked the audience the following question... "How many people in the room have electricity in their house? Hands up please". Everyone put their hands up of course with some bemusement. I then preceded to ask another follow-up question "How many people have a gas pump in their house?" I could see they got it quickly.

Access to home charging makes the use case for electric completely different to the internal combustion car. With a tiny bit of planning, in advance of any long journey, you will fully charge your car at your home. Not everyone has access to a charging point at their home, for example in apartments, and I will deal with this and the solutions in the next chapter on charging. You don't even need to install some expensive home charging wall box in your driveway or at the front of your house. These typically cost between 600 to 1,200 euros, around the same in US dollars.

I have been driving an electric car for over 5 years and still haven't got around to installing a wall charger at my home. I usually just plug my car into a normal 220-volt domestic socket once or twice a week. When plugged into this normal wall socket, the car charges slowly, between 12 and 14 kilometres (7-8 miles) an hour of range added to the battery. This may seem extremely slow, however cars are unused 90% of the time. Parking my car up usually around 8 pm in the evening, I'm not leaving my house till 8:00 am the next morning which adds around 150 kilometres (100 miles) of range overnight. I could install a home charger and fully charge my car in around three hours, so far, my 220v normal plug is enough for my average daily commute, which is usually around 25 to 30 kilometres (15-18miles). When I plan a long journey, I usually plug the car in two nights in a row before I leave, which gives me a starting range of around 380 kilometres. If I had a gas pump in my home, I would fill the car up before I headed out on a long-distance journey.

As with a gas car, you never really get the stated range. I've got enough speeding tickets to tell you I drive too fast, and I have never managed to get 380 kilometres (230 miles) of range out of my Tesla Model S. Driving fast eats range. Also, on extremely cold days, when it's below zero Celsius (32 Fahrenheit), the range further diminishes. I would estimate I normally get about 250 kilometres range out of my car from full to zero, with fast highway driving. Electric car range is measured as Miles per Gallon equivalent (MPGe) in the US. This is calculated by running the car on a dynamo from nearly a full capacity battery to zero. This gives a falsely high range number as it doesn't represent real-world conditions. For example, the Tesla Model Y has shown 451 miles of range. The US EPA then applies an arbitrary .7 multiplier to determine a more realistic range, in the case of the Model Y that comes out at 361 miles of range. In Europe, the range tests are known as the New European Driving Cycle (NEDC), and Worldwide Harmonised Light Vehicle Test Procedure (WLTP). The newer WLTP test claims to provide estimates within 10% of real-world experience, my experience is more like 20%.

Manufacturers often overstate the range capabilities of their EVs. Regardless of the claimed and rated range of an electric car today, you can forget about any of the manufacturers' stated range. Unless you are driving like in the movie Driving Miss Daisy, you will never, ever hit this, just like you probably never hit it with your gas car. I used to do a long-distance electric drive, before the pandemic, at least once every two weeks, I have done Luxembourg to Brussels and back drive hundreds of times. I've also driven from Luxembourg to the Netherlands and back in the same day (around an 800km / 500-mile drive) and it works. Having analysed thousands of our customer journeys, not one achieved a rated or manufacturers advertised range, with an average deficit of 10-22%, depending on the time of the year. You should be aware of this, don't let it put you off. There are already enough fast chargers across continental Europe to cover pretty much any journey today. The picture in the US is slightly behind, however, rapidly improving. Other parts of the world are less developed, and I will cover this in the next chapter.

Before I get onto the subject of how much range is enough for the average driver, I'd like to tell you a story about my first very long-distance drive in an electric car. I had just left my corporate life to begin working on our start-up and bought my brand-new Tesla Model S. I collected the car on a Saturday and drove it home to my house. I drove it around for a couple of days and I can tell you I rarely stopped looking at the range indicator on the little battery icon on the dashboard. I too was brainwashed with the fear of 'range anxiety' and couldn't take my eyes off the kilometres remaining indicator. I would look at road signs that said four or five kilometres to my destination then, knowing the miles remaining on my battery, try to calculate when arrived, just how accurate the range predictor was. Even though the Tesla has an energy monitor feature that can show you the predictability of your range I still checked it repeatedly.

There was a reason for this, firstly it was my first time owning an electric car, and secondly, one week after I bought the car, I had a scheduled, ultra-long journey planned. For years I and my partner talked about getting a dog and we had settled on a Labrador retriever. We had been looking for months to try and get a Labrador puppy in and around the Luxembourg region. The only breeder we could find that was planning a litter was in Trier in Germany, just across the border from Luxembourg, they weren't planning to have the pups for another six months. We were tired of waiting. Through a breeder friend in Ireland, we contacted a Labrador retriever puppy farm in Slovakia. We called the lady that ran the farm and she told us she had a Golden Labrador retriever puppy, the last one remaining from a litter of 6, available. We decided we would drive from Luxembourg to collect the puppy the following weekend. It is almost 1,200 kilometres door to door, around 750 miles. Although we planned to stop on the way and stay overnight in Germany, we decided to do the drive back in one day. This trip was planned one week after I got my first electric car. I told my partner there was no way I was going to attempt such a journey in an electric car as I had no experience and had never even driven that much in a gas car in one day. I even looked to rent a diesel car for the trip.

With visions of being trapped in mountains with snow and zero battery (the trip was the last week in November), I decided to go for it. Planning, that's what was necessary for a trip like this. While I was worried about navigating such a long distance in one day in an electric car, I was equally worried about the fatigue, and whether I would be dangerously exhausted towards the end of the drive, having never driven this long before in a single day. I'm sure many people reading this have done longer distances, when you're from Ireland, a 1,200-kilometre drive is pretty much traversing the entire country several times, I wasn't used to it. The planning took me just a couple of minutes. By simply typing in the address in Slovakia, the Tesla worked out the four charging stops I would need to get there. These were all fast chargers on the Tesla network, with speeds of up to 150 kW. Charges of this speed can charge a 75 kW Tesla battery from zero to around 80% in around 35 minutes. I'll explain all of this in a lot more detail in the next chapter. The Tesla screen interface only displayed the Tesla superchargers on the route. With a little bit of internet searching and playing around online with charging maps, I was also able to select other superchargers that were non-Tesla, for example, the Ionity network. All of this might sound like a lot of work when you're used to just driving and filling up with gas, but this was a one-off mega-long drive and my first time doing something like this in an electric car.

I like to stop on long road trips, at those mega highway super stops, where you can get nice coffee, use the bathroom, and freshen up. I was really surprised to discover that some of the scheduled Tesla stops, had their own rest facilities. You are never really driving more than three electric hours before you must charge. I was fresh after the first 250 kilometres and didn't want to stop. I could have gone much further before I needed my first stop, but I was forced to stop to charge the car. The first stop took around 40 minutes to charge the car up, I was a little frustrated because I wouldn't have stopped for that long having only covered less than 300 kilometres so far, however, we had some lunch, freshened up and resumed our trip. Charging stop number two was at around the 500 kilometres mark and charging stops three and four, at around the 700 and 1,000-kilometer mark, these were very welcome. I began to look forward to the next charging stop where I could rest and refresh myself. On the last charging stop, I even had a little nap for about 25 minutes.

We both thought this was kind of a crazy trip that could be dangerous with excessive fatigue towards the end of the journey; I even had visions of arriving at the destination and collapsing with exhaustion having driven for over 12 hours. As it transpired, after leaving Slovakia at 7:00 AM for our return journey, we drove into our house in Luxembourg at 10:00 PM and I was not exhausted. I have been much more tired driving a gas car on a 500+ kilometre journey in the past because I didn't stop. The forced charging breaks, where you get at least half an hour rest from driving, makes a huge difference. This trip taught me something, it was the first time I realized that driving a gas car was completely different to driving an electric car, and that range limitation for very long journeys doesn't exist. It's just different. While the charging stops made the trip two hours longer, they were worth it. My 2,400-kilometer electric odyssey was a baptism of fire, I had survived an ultra-long distance EV drive and learned so much and it's one of the reasons why I could never imagine going back to a gas car.

If you live in Germany, the average number of kilometres you drive per day is just 37. In the Netherlands, that number is 35, in the UK it's 34 while in Italy it's as low as 25 on average driven per day. In almost every European country, the average amount driven per day has gradually declined over the past 20 years. In the UK, for example, back in the year 2000 the average driven was 41, so that went down 7 kilometres by 2020. There are multiple reasons for this gradual decline, however, it is in part due to changing ownership habits for cars, greater options for mobility like ride-hailing, sharing and rental. Undoubtedly the massive shift to work from home due to the pandemic will further drive these averages down further. Regardless of the reason, these numbers are well within the range of every electric car available. In Europe's largest economy, Germany, just 1% of daily driving commutes last 120 minutes or more, with most journeys between 15 minutes and one hour.

In the USA the average daily drive is just 25.9 miles per day, (41 kilometres). As in Europe, American drivers are driving far less than they were back in the year 2000 and there has been a gradual decline in most states. If you ask yourself just how many kilometres or miles you do per day, it's likely you're somewhere around these averages. This data will of course have outliers, there will be people who do much longer commutes, and some much shorter, but these will be the exception. I once worked with a guy who drove almost 200 kilometres every day to work and 200 back, but these are the rare ones.

So, if most of us drive less than 50 kilometres per day, why is there an obsession with the range of electric cars when most electric cars have at least a 300 kilometres range?

It would seem irrational to have range anxiety with such small daily commutes. This is the very crux of the range debate; for 90% of the time, we use a car, and electric is perfectly workable. Range anxiety fear comes from that 10% of driving when we plan to go beyond the range of the car. I can tell you that whenever I speak to anyone about long-distance driving in an electric car today, they will bring up the time they drove 400 - 500 miles in a single day and how this would never work in an electric car. However, in analysing long-distance journey patterns, 95% of those long journeys will be on highways or major routes. In Europe, and most major highways in the US, highways are now well serviced with charging and supercharging stations for electric cars. I'll cover this in the next chapter.

The issue with range anxiety today is one of irrational human fear.

We conjure up an image in our head, just like I did before my first big EV road trip of running out of charge in the middle of nowhere. Now I'm not saying this can't happen, it can and did happen to me, as I'll explain below, however, it is an ultra-rare event. Nonetheless, unless range anxiety is banished, people will continue to be slow to adopt electric cars. This is where education and most importantly ease of use is paramount. Ease of use is vital for everybody to be comfortable enough to make the switch to an electric car.

I mentioned before that 90% of the time electric cars work for me, but 10% of the time they don't. I have become frustrated on a few occasions with the limitations of an electric car, however, thankfully these are very rare and usually my fault. Unexpected long journeys are the issue for me. On a couple of occasions, I received short notice to do a trip. This happened to me in 2019; I was working late in my office when I got a call to attend a very important meeting first thing in the morning in Rotterdam. The drive from Luxembourg to Rotterdam is about 300 kilometres but my car only had about 30 kilometres of range left on the battery (I had planned to charge it that night). I was tired and wasn't looking forward to a long drive and to make it worse now I had to supercharge on the way because I wouldn't have enough range to make it. So instead of a two and a half-hour drive my drive was three hours because I wasn't charged up in advance. This is the only limitation I've experienced in over five years of driving an electric car. 90% of my daily driving is covered, and what's even more convenient is I don't have to ever go to a petrol station, which is time-saving. When I must do a long journey, I do a little bit of planning and it always works out. With an ever-increasing number of supercharging options, same day long-distance return journeys are more than possible. Electric cars work, today, and for most of your daily car needs; it's just a little different and requires some small habit changes. It's also important to remember that even though it works today, range is only going to become a smaller and smaller issue as charging availability, and fast charging networks, grow exponentially across the world.

However, before finishing this section of the book, I'd like to tell you a story about the time I ran out of power. This has only happened to me once, and was mostly my fault, but also highlights why ease of use is fundamental to charging. Ease of use is more important than the range of your car's battery, and the number of chargers available. I was lucky enough to borrow a Tesla Model X (the seven-seater) that I needed for a New Year's Eve trip with children and a dog. We were driving from Luxembourg to a holiday village in France. The weather was below freezing, minus 10 degrees Celsius, for much of the drive. The car was super heavy with seven passengers, a dog and luggage. The kids were using their heated seats, charging their smartphones and I was driving, as usual, a bit faster than I should have been.

The expected range of the car should have been just enough to make it to our destination, around 280 kilometres away. However, with all the power drain and cold weather, it quickly became apparent that we weren't going to make it on a single charge. The Tesla navigation system advised me to supercharge for 15 minutes, about 70 kilometres away from our destination. The car informed me we would have around 20 kilometres of range remaining when I got to this charging station, so there was nothing to worry about. This Tesla charging station was on the main highway between Paris and Reims in France. It was extremely cold, and the battery drained to the point where I began to slow down even before I got there. As we approached the turn off for the gas station, where the chargers were, the car was down to just 12 kilometres remaining, but again I wasn't worried, we were almost there.

I took the turn for the gas station, as indicated by the Tesla navigation system, it was one of those huge highway service stops for trucks and cars, complete with all the usual facilities. The SATNAV said I had arrived at my destination, however, we couldn't see the Tesla superchargers. There was no simple sign with an arrow pointing toward the supercharger's exact location (this is still a problem with most Tesla supercharger stops that I'll talk about later). As we drove in, I immediately had to make a lane choice between parking, the gas pumps, and the truck stop. These lanes were divided by concrete barriers, so with other cars behind me, I had to quickly choose a lane. I wasn't driving a truck and didn't need gas, so it was logical to choose the lane for the car park where I assumed the charges would be – right?

Wrong! Almost the instant I chose the car park lane, my son, sitting in the back seat, said "I see them, they are to the right". Too late. Once you drive into the car park there is no way to turn back, it's a one-way concrete divided system, and the only way to get back to where you were was to go back onto the highway and find a U-turn opportunity. With a line of cars behind me, I had no choice but to drive back out onto the highway. Now with panic creeping in, I re-entered the charging stop back in the navigation system to find that it was a 20-kilometre round trip (as the nearest exit and re-entry point on the highway was 10 kilometres away). I had just 11 kilometres left on the battery, so staying calm, I drove slowly towards the next highway exit, I decided it wasn't a good idea to break down in the cold on the side of the highway and it would be better if I get off the highway as soon as possible.

I took the first exit off the highway, which didn't have a return to the other side. I stopped the car on the side of a small road and called Tesla support. They gave me two options, wait for a tow truck or try and make it to another supercharging station which was just 15 kilometres away from the highway. I decided to leave most of the kids and my partner at a cafe stop before I tried to make it to this supercharger rather than have them stuck in the car. I drove extremely slowly (this is called 'hypermiling' in electric car terms); if you drive slowly the battery drain is not as extreme and you get a little bit more range. With still 9 kilometres left to reach the charging station, my car went to zero. We kept moving, just like in gas cars we had a bit more left in the tank so to speak even though it was saying zero. At this stage, I was literally in the middle of nowhere driving across a flat swamp-like landscape on a small country road when the car just began to slow down. No matter how much I pressed the accelerator it was clear this was the end. The big Tesla just ground to a halt and then, in a final dramatic step, both the dashboard screen and large centre console went black. The electric windows still worked, because they are run by a separate small 12 Volt battery, just like an internal combustion car, for emergency purposes, but otherwise, the car was dead, dead. I had to wait for a tow truck, which was a whole other story trying to explain in my bad French why one couldn't drag a Tesla up the ramp of a tow truck, however, in the end, after about an hour we ended up at the charging station.

The whole debacle lost us two hours and highlighted just how wrong it can go in an electric car. In this case, nothing went wrong, the car brought us to where it was supposed to bring us. A simple $50 sign with an arrow pointing to where the chargers are would have saved all this hassle. Having used superchargers across Europe since these signs still don't exist in most cases. I have often driven into a supercharging destination, sometimes in a hotel complex, to have to drive around before I find the chargers. Many people have said the same, so I even wrote to Tesla and told them to invest in simple direction signs. This story is about electric car range and user error, as in me, but also about making the user experience simple and easy. I will explain in the next chapter how sometimes overly complex, brand new charging stations are being installed right now with multiple different systems for charging, paying and navigating. I liken the global charging landscape to the early days of the Wild West and will explain why later in the next chapter.

I'm going to summarize the key facts and fiction in each of these sections as follows:

FICTION: Electric cars do not contain enough range today for my everyday needs. EVs are far more difficult to use than gas cars and are just not practical yet. There is a high possibility of running out of battery in the middle of nowhere and becoming stuck. It's not possible to do more than 500 miles of driving in a day. Electric cars will never replace gas cars for people who need to drive long distances every day.

FACT: Almost every electric car on the market today can deal with 90% of your driving needs. Most journeys are well within the range of electric cars today. For long-distance driving in much of the US and Europe, highway supercharging networks are already sufficient for even the longest drive. It takes a different approach, a change of old habits, and a small bit of preparation to make sure you don't get into trouble on long-distance electric driving. If you have access to even a standard 220 Volt socket at your home, you will probably only ever use chargers on the occasional long journey. When your range management goes wrong, it can go wrong resulting in a potential tow truck event, but this will be almost as rare as running out of gas.

NOWHERE TO CHARGE

While the range of electric cars today is more than sufficient in most cases, having access to chargers, knowing where they are, and ease of use is vital. The second most common myth uttered by the uninitiated is that there is nowhere to charge. The day before I wrote this section of the book I was out cycling in Ireland, and I stopped for a cup of coffee at a rest stop in North County Dublin. I got talking to two older guys who were also out for an afternoon spin. The subject of what we do came up, and I told them what I was doing with electric cars. The instant I told him about my job, they immediately said “this country is terrible, there’s nowhere to charge electric cars”. This is not a comment unique to Ireland, I get this every day. The most common comments I hear are “there is nowhere to charge”, “electric cars can’t go mainstream for years until there are thousands of chargers installed”, “every gas station needs to be converted to electric before people will be able to charge”, “there are not enough chargers and when you get to them, they’re always occupied so you have to wait hours”.

These are just an example of the comments I hear repeatedly when the subject of charging comes up, usually made by people who have never even driven an electric car before. To address the older gentleman’s comments, in Ireland there are over 1100 public charging stations across that small country. This doesn’t include destination chargers, which are chargers at hotels, restaurants, shopping centres, and supermarkets etc. There are lots of chargers in Ireland, that’s not the issue. The question is firstly where they are and secondly, and more importantly, are they working and available. Connectivity between the car, the driver and the charger are more important than the number of chargers. I’ll explain in the next chapter why this is the case, and how Tesla has gotten this right while many others have not.

The perception is that there simply are not enough chargers for electric journeys today, thereby providing an excuse not to choose an electric car. This is not true. I’m not saying that there are enough chargers going forward, that's different, we need thousands and thousands more chargers in Europe and the US to deal with the rapid mass adoption of electric cars. As far as today goes, there are already enough chargers for almost every journey use case you could imagine, unless you plan to go off the grid and deep into a wilderness somewhere. The problem I see, from my own experience and from the thousands of electric journeys our customers have made, is not the number of chargers, but how easy it is to find them and use them. Knowing if they work, are available and if the payment for the charging is integrated and simple. This isn't the case at all today which makes electric car adoption more difficult and gives some truth to the statement that electric cars are not easy. I'll explain all of this in the next chapter.

In the US there are over 48,000 public charging stations containing around 100,000 charge outlets across the country. There is already almost half the number of charging stations as there are gas stations in the USA. In Europe, the number of charging stations is higher, with around 250,000 charging points. This number grew from just 2,500 stations a decade ago. Globally, as of 2020, there are over one million charging points across the world. These numbers are growing exponentially. The European Union has set a target for one million operational charging points by 2024. Within five years there will be more charging stations and more charging outlets than there are gas stations and gas pumps in the USA, European Union, and the UK.

Terminology is important here as it's not that easy to find out just how many charging stations and individual charging points (gas pumps) exists. Often the word 'charging point' and 'charge station' are interchanged, both meaning different things. For this book, and most references, a charging station is to be regarded just like a gas station which can contain multiple charging points. China has over 800,000 charging points, having installed over 500,000 in 2020 alone. China is installing over 1,000 EV charging points every day. You may be wondering where all this electricity will come from when everybody is driving an EV.

To find all these chargers, there are lots of charging apps and online charging maps. Examples of charging apps include ChargePoint, PlugSurfing, Chargehub, Zapmap and now even Google Maps can show you where the chargers are. The big problem is these are often community monitored, meaning people update the status of a charger based on their last experience. Not all charging apps and charging maps are the same, some show you different networks while others exclude some networks. Some apps have greater connectivity in real-time to chargers while some don't. Although the world of EVs is still relatively new, many of these chargers were installed 10 years ago and do not have any connectivity, these are dumb chargers. What's worse is many of these chargers need pre-registration, credit card type swipe cards, and have multiple different payment systems. Hence why I refer to it as the Wild West.

Even with all the charging infrastructure being built out today, even with all the public charge points in hotels, restaurants and supermarkets, there is still always the last resort option of plugging your charging cable through a window and into a normal 120/220-volt socket if you need to. Remember this option doesn't exist for gas cars unless everyone you know keeps a spare can of gas at their home, or you want to siphon some of their petrol. Indeed, the concept of siphoning is also about to become mainstream for EVs as new car models will offer the option to share charge. Many car manufacturers are planning to offer this feature, and some already do. The solar car I mentioned in Chapter 5, the Sion from Sono Motors, can do this.

The number of chargers is not the issue, making them as easy, or even easier than gas pumps to use is the issue. I am astonished how many times, in this 21st century, the user experience is badly considered. The world has become busier, and we are all bombarded with endless information, apps and features. What we need is hassle-free, time-saving user experiences, not new technology or services that consume more of your time. Sadly, many of the charge point operators, apps and electric car manufacturers are truly missing this point. I'll explain my view and how this should be made even simpler than using a gas pump later in Chapter 7.

FICTION: There is no widespread charging network in place. Chargers are hard to find and don't always work. Most countries haven't put enough chargers in place for the number of electric cars on the road today. It's dangerous to attempt a long-distance journey currently due to a lack of chargers.

FACT: There are already enough chargers for most of your driving needs. Forget about how many chargers there are or aren't; you don't know how many gas stations there are, and you don't need to know how many chargers there, this is no longer important. With the rate of buildout now taking place it's going to become a non-issue very quickly. Unless you plan to go way off the grid, you will be able to find a fast charger for your journey. The real issue is the ease of use and connectivity to those chargers, and this needs some work.

AFFORDABILITY

Today the perception is that most electric cars are still more expensive than their gas equivalent sticker price. This is not true in many countries. Although there is nowhere near the range of models for electric yet, there are some like for like equivalent size and power cars where indeed the cost of electric is lower than gas. However, to fully measure the price difference between electric cars and gas cars, we must compare the entire lifetime cost. Servicing and maintenance, as well as the difference between gasoline prices and electricity costs, need to be considered. Resale values also need to be considered, as this is still evolving for electric and hasn't matured with the same predictability you get with gas cars. Then there are other dynamics at play, for example, many countries and U.S. states offer incentives to buy electric cars. This is one of the main reasons why Norway now sells more electric cars than gas cars. Governments have a place to play in incentivizing the uptake, however, subsidies can't last forever, they're good just to get the ball rolling. What must happen is that battery costs must continue to drop to the point where government subsidies to buy an electric car are no longer necessary.

The question of affordability isn't a simple or black and white question, because you need to compare the cost of gasoline versus the cost of electricity. In Europe for example, the cost of petrol and diesel is far higher than in the US. While in the Middle East the cost of gas is extraordinarily low. In early 2021 the average cost per gallon of gas in the United States was $3.19 ($0.84 per litre). In Europe the average cost varies a lot, for example in the Netherlands it was €1.85 ($2.18), in March 2021, while in Bulgaria it was just €1.05 ($1.24). This means an American living in The Netherlands would pay over $8.00 for a gallon of gas.

Gas is extraordinarily cheap, cheaper than orange juice and bottled water in some cases; it is also subsidized in many countries, and I'll explain how the great energy transition will change this in Chapter 10. For now, when comparing the cost over the lifetime of an electric car versus a gas car I'm going to just assume United States gas prices, even if they are artificially low. One of the other issues in comparing the cost of gas versus electric cars is that most manufacturers have started at the top with luxury models first. These high-priced cars are often €10,000 higher for electric than their gas equivalent, however, that needs to be taken into the context of the total cost of those cars. While paying €10,000 more for an electric car seems a lot more, some of these cars cost as much as €100,000, so it's just 10% more in those cases. Paying 10% more for a cheaper to run car may be worth it over the lifetime of that vehicle.

Before we get on to the absolute cost of a new car let's look at the lifetime costs first. A 2018 study from the University of Michigan's Transportation Research Institute found that electric vehicles cost less than half as much to operate as gas-powered cars at an average of $487 per year to run, versus $1,117 for a gas car. Parts need replacing in gas cars, the bigger and more complex the part, the more the cost. With electric cars, having far fewer moving parts means they require massively less maintenance. Tires do wear out faster in an electric car because of the greater amount of torque and instant acceleration being applied to the wheels, however, thanks to regenerative braking, brakes don't. Tire manufacturers are developing specific tires designed for electric cars.

If you plan to keep your electric car for just three to five years, there's a very good chance you will spend nothing on servicing and maintenance. Depending on how much mileage you do, and where you live, typically the cost per mile for electricity is 38% of the gasoline-equivalent. The average cost per mile for a gasoline-powered car is 9.3 cents in the United States versus 3.6 cents for an electric car. These numbers will of course vary considerably depending on where you live, if you live in Saudi Arabia, you will pay just $0.91 for a gallon of gas, that's just 23 cents per litre, 34 times less expensive than if you lived in the Netherlands.

It's not just the gas price savings or the lack of lifetime maintenance cost that makes an electric car potentially far more attractive - they may last virtually forever! The electric car world is new, gas cars have been around for over 100 years and typically they run for around 13 years. It's too soon to see how long an electric car will last. Sure, eventually the battery will degrade, and right now most electric cars can't have their batteries easily replaced, but apart from that, there's not much else that can wear out. In theory, you could drive your electric car for decades. This means all the traditional depreciation models we have for gas cars don't apply yet to electric cars. This has a big impact on long term resale values which I cover in the next section.

A Tesla Model 3 standard range starts at around $37,000 in California. The fully loaded high spec version could be as high as almost $68,000. These cars are subject to a $2,000 or even $4,500 state rebate (based on income eligibility). In Germany, the rebate is €3,000 ($3,300) for a Model 3, while in Ireland it's as high as €5,000. In the UK new electric cars below a certain price are eligible for the plug-in car grant (PiCG) of £2,500 ($3,500). However, it is very likely you cannot rely on these incentives forever as electric cars become more mainstream. There are other benefits than just rebates, for example in certain countries annual road tax is low or even zero for electric cars. In other countries, parking charges are lower and city congestion charges, as in London, is zero for electric cars while it is £12.50 for a gas car to enter London's congestion zone.

In the United States, a standard 1.4-litre VW Golf starts at around $23,000 while its newer electric cousin, the ID 3 costs over $40,000. Even with government rebates, fuel, and maintenance savings over three to five years, the gas Golf is more affordable, however, these are not like for like comparisons. The Volkswagen ID3 is still not really in the same segment as a Golf, although it looks very similar. It is more equivalent to a high performance or premium edition of the Golf which can cost well over $40,000.

I am deliberately not giving you a side-by-side table style comparison of a gas car versus an electrical equivalent, because it's hard to pick a fair like for like car model. The performance of electric cars, in terms of horsepower and acceleration, for example, is far superior to most of their equivalent sized gas cars. However, in the mid-segment and luxury segments, today, most electric cars are 10% to even 20% more expensive on the sticker price, even with government rebates. When you consider the lifetime savings of an electric car versus gas this increase reduces to around 8% to 15% over a gas car. You could argue that paying 8% to 15% more for the benefits of driving an electric car and the air quality and climate impact is worth it, however, unfortunately for most people paying this green premium isn't within their budget just yet. Parity for gas versus electric pricing is getting very close. In some countries, with rebates and higher gas prices, electric cars already are on par with the gasoline-equivalent over their lifetime. As battery technology improves, and economies of scale begin to kick, we will near the point, probably by the end of 2022, where an electric car is more affordable over its lifetime than the gas equivalent. Research firms and industry followers sometimes claim this has already happened, and that is the case in some certain markets, but most analysis points to between 2022 and 2025 at which EVs become more affordable than gas. For example, BloombergNEF's analysis predicts lithium-ion battery costs will fall to the extent that electric cars will match the price of petrol and diesel cars by 2023 and McKinsey's Global Energy Perspective 2021, forecasts that "electric vehicles are likely to become the most economic choice in the next five years in many parts of the world".

Breaking the $100 per kWh for battery cost is the threshold by which EVs would become cheaper. As covered in previous chapters, the race is on to produce a cheap electric car. Volkswagen, Tesla, and others are close to unveiling a car costing $20,000 or less that has sufficient range for everyday use and practicality.

FICTION: Electric cars are more expensive to buy and own than gas cars. Electric cars are years away from being cheaper than the gas car equivalent.

FACT: To properly compare the cost of a gas car versus an electric car there are many variables, including where you live, how much you pay for gas, how much you pay for electricity and what government incentives are available. On average an electric car will cost at least 10% more than the gas equivalent, however over life they now get close to parity. We are less than two to three years away from the point by which electric cars are cheaper to buy and own than gas cars.

RESALE VALUES

The cost of a used car is determined by mileage, the condition, the model of the car and what country you live in. The average American keeps their new car for around 8.5 years while for the average European it's 11.5 years. Like gas prices, this varies wildly depending on the country you live in, and the wealth of that country. For example, in Luxembourg, the average car ownership length is just 6.5 years while in Eastern European countries, with lower GDP per capita, that can go as old as 16 years. Worldwide the average is 12 years.

You can pretty much predict, with reasonable certainty, what your used car will be worth in a couple of years based on mileage and condition. These depreciation models have been honed over decades in the used car trade. This is not the case for electric cars. The average mileage on a car being scrapped is 144,000 miles (230,000 km). You have probably heard your taxi driver tell you how they've done a quarter of a million miles or more in the same car, however, these are the outliers and they have likely replaced several major components like gearboxes already.

How long electric cars will last depend mostly on the battery's lifespan, which I'll cover in the next section. I'm going to base this section on the real-world resale values being achieved for electric cars now, in 2022. Gas cars depreciate between 15% and 25% per year, with the biggest hit being in the earlier years; typically, after five years a car has lost 60% of its value.

In the past, electric cars were depreciating at a faster rate than gas cars, at around 52% in the first three years compared to 39% for gas cars. This has now changed considerably and for example, the Tesla Model 3 has reversed the trend of vehicle depreciation and outperforms gas car equivalents. In the UK, as of 2022, EVs retain almost 49% of their value after three years compared to just 40% for a gas car. Supply and demand also play a factor here. In many countries, there are waiting lists for new EVs. This artificially inflates the used car cost for an electric vehicle; this will be temporary of course, as supply begins to meet demand, although the Covid pandemic and chip supply shortages are distorting values even more.

The biggest challenge for the auto sales industry is in trying to predict what will happen now that the tipping point has been reached for electric car adoption. You can no longer use the predictable linear models for depreciation with gas cars because something huge is about to change - demand. I've covered all the regulations that will ban the sale of new gas cars from around 2025 onwards, plus consumer awareness of their air quality and climate is growing. There are other factors at play, for example, there is an increasing shift away from individual ownership to shared fleets. As global fleet sales increase and individual sales decrease, the global players will be playing big number economics. Financing thousands of vehicles come under much more sophisticated financial due diligence than a single car purchase. The transaction size is considerable, meaning the assets underlying the loans as collateral need to be assessed. Leasing companies and bankers finance depreciating assets using predictable models. If these models no longer hold, say from 2024 onwards, how can you propose financing terms with certainty of return to fleet providers for electric? This is a problem I have become aware of and have indeed been in discussions with several leasing companies and finance houses over the past number of years. I can tell you the vast majority of some of the biggest leasing companies in the world haven't yet figured this out, as of early 2021, because there simply isn't enough historical data on the resale of electric cars over a longer period.

All other things being equal, it is likely that demand for new gas cars will fall off a cliff from 2024 onwards. Why 2024? because if you assume you will hold your new car for seven years, and you purchased it in 2024, you're going to sell it in 2031. By 2031 multiple U.S. states and European countries will have banned the sale of new internal combustion cars, no one will be interested in your old petrol or diesel car. This combined with the fact that battery costs will have continued to drop, and a new electric car will be cheaper than gas, why would you want an old gas car? Price. There is always a market for cheap goods. It is likely that by 2031 the only way you will offload your old gas car will be to sell it at a considerable discount, even beyond the normal depreciation we are used to. This fact has not been lost on the big leasing and financing companies. They have told me this in various meetings and I have witnessed a more relaxed attitude towards financing EVs and a more concerned attitude to longer-term financing of gas cars.

This is a very important point. In comparing the resale value of an electric car, you purchased today, you need to compare it to the resale value of a gas car you purchased today but under accelerated depreciation rules. To be conservative you can assume depreciation will be at least 10% worse than it normally is for gas cars from 2023 onwards. This has already happened in many model examples. Let's take a Mercedes-Benz E Class, one of the mainstays of Mercedes with an excellent resale value historically. After three years a typical E Class will lose around 47% of its value in the US. Compare this to a Tesla Model S, which loses just 36% of its value after three years. While the Tesla Model S is more expensive to purchase than a Mercedes E Class, after three or even five years you recoup the money.

Lower down the scale, the sturdy reliable Volkswagen Golf will still be worth around 67% of its value after three years currently, while a Tesla Model 3 is estimated to be still worth around 60% of its value after three years; these numbers assume 12,000 miles per year. However, these estimates are just projections in the case of Model 3, as supply and demand play a big factor here. Demand for electric will increase while demand for gas will reduce.

FICTION: Gas cars hold their value better over their lifetime than an electric equivalent. With the ever-increasing advancements in technology, electric cars quickly become out of date, thereby destroying your car's resale value.

FACT: Electric cars and gas cars are on parity in terms of depreciation in some cases, but in most cases, electric cars now hold their value better than gas cars. Although there are great advancements taking place in battery technology, this is not having a massive impact on future resale values yet. A breakthrough could change this of course. When considering buying an electric car you need to not just compare the depreciation rates of gas cars today but the likely depreciation rates of gas cars soon as they begin to phase out and electric cars scale up. As of 2022 gas cars retain around 40% of new car value after three years while EVs retain 49% of their value after three years.

BATTERY LIFESPAN

"The battery won't last very long and you're going to have to replace it before you sell your car." This is one of the most common things I have heard about electric cars. Old lead-acid batteries die eventually. Rechargeable Nickel-metal hydride (NiMh) and nickel-cadmium cell (NiCd) batteries die eventually. These are the most common AA and AAA cells you use in everyday electronics. Today's more modern rechargeable lithium-ion batteries also die, eventually. In Chapter 8 I will explain the great advancements taking place in battery capacity, charging speed and longevity. The purpose of this section of the book is to dispel the myth about the short lifespans of electric car batteries. Batteries are in the business of moving ions from one point, the cathode to another, the anode. This movement is what generates the electrical charge. Constant charging and discharging over time gradually reduce battery capacity and maximum discharge power. This is a chemistry problem.

In new healthy batteries, the ions flow freely between the cathode and the anode. A lithium polymer battery can lose up to 20% of its capacity after around 1,000 charging cycles, this is common for a smartphone or laptop today. Batteries can even degrade if they're not used, and temperature also plays a big part. How a lithium-ion battery is used daily also determines its capacity and lifespan. In a lithium-ion battery, crystals form over time, called dendrites. These dendrites build up like stalagmites and stalactites in a cave and the cumulative effect gradually reduces capacity and battery performance. Batteries also contain a charging "memory" which can affect its longevity, I'll explain the science of this later.

The problem is most of today's statements about electric car batteries are based on knowledge of old tech lead-acid and nickel-metal hydride batteries. Modern lithium-ion batteries last far longer than their ancestors. Lithium-ion batteries perform better than predecessors with rapid charging and discharging and were designed specifically for this purpose. When you buy a gas car, you usually get a warranty for three, five or seven years and even longer. Electric cars also come with warranties. Just looking at Tesla as an example, the capacity of the batteries is guaranteed up to 70% for eight years or 100,000 miles or even 150,000 miles depending on the model.

The maximum range possible for a lithium-ion battery does gradually reduce as it ages. Multiple models from real-world data and simulations show, and can predict, just how much range will be lost from the battery over its lifetime. There is a mistruth, and I've often heard outrageous statements, that batteries will lose 20%-30% of their capacity within just two years. This is not true. Tesla batteries, for a Tesla Model S or X, have been run out past 200,000 miles, with at least 90% plus of original range still capable. The reason batteries lose range over time is both chemical and due to many internal short circuits as some cells die. This should not be a cause for concern as 200,000 miles is by far an outlier in normal car usage. On average, and subject to use and driving style, a Tesla battery pack will lose less than 10% of its capacity after 160,000 miles. A Tesla battery with 300,000 to 500,000 miles, over its lifetime, can then still be reused for purposes other than moving a car. I'll explain the second life of electric car batteries later in the book.

Electric car batteries could even last for 20 years based on current technology longevity forecasting. The million-mile battery has also been forecasted as technically feasible soon. When I purchased my Tesla Model S with a 75-kW battery, fully charged it displayed 390 kilometres of range, brand new. The maximum range estimator does change from time to time, as it is based on your driving style and therefore updates predictability accuracy. There is also a sweet spot to charge a battery between 20% and 80%, where it never fully drains and is never repeatedly overcharged as this is bad in the long run. After 60,000 kilometres, and four years of driving, my Tesla, when fully charged now indicates 375 kilometres range and sometimes even shows as high as 380 Kms. This is just a 3% to 5% loss of range after 60,000 kilometres.

I am only choosing Tesla here as the example, however, many other manufacturers batteries perform similarly. Older edition electric cars, with the older generation, lithium-ion batteries used to degrade faster, but this has changed considerably in recent years. Loss of efficiency, loss of range and performance is not unique to electric cars. Diesel cars for example lose efficiency and range performance over their lifetime. This is because components wear, injectors become clogged with carbon burn residues, cooling systems become a little less efficient and generally, components like pistons simply wear down. This efficiency loss needs to be added to the already 80% loss of energy from an internal combustion car regardless of it is new or not.

About replacing your battery, this is an old story, from the very earliest era of electric cars before modern battery management systems. There's been a lot of debate and new start-ups looking at battery swap technology, which I will elaborate on later. For this section, I'll answer the question of whether you'll ever need to replace the battery in your new electric car. The answer is no. The main reason for this is that for almost all new electric car models on the market today, it's not even possible to replace the battery. The battery pack is integrated into the car's floor plate foundation and not removable until the car is eventually being scrapped. Indeed, integration of the battery into the floor plate is increasing with the latest generation battery packs by Tesla and Volkswagen. This will make it effectively impossible to ever replace the battery during the life of your vehicle. Unless you have a defective battery pack, in which case the manufacturer will likely replace your entire car under warranty, you will never see, maintain, or change your battery.

To compensate for capacity lost over time, electric vehicle battery makers ensure there is some 'buffer' space available to compensate for range loss. So as the vehicle ages, additional spare capacity is made available. In the case of Tesla, over the air software updates can even release additional battery capacity range. This was well publicised during Hurricane Irma in Florida when in 2017, owners of some Teslas received additional capacity, as much as 40% in some cases, that Tesla released to them wirelessly to help them evade the hurricane. These were in effect free software upgrades releasing more battery capacity than the owner paid for when they purchased the vehicle.

Unless you're doing something extreme with your car, like fully charging and fully discharging your battery every single day, and overusing superchargers, your new car battery will only lose around 10% during the life of your ownership. Battery life span has been proven to have little or no effect on current electric car resale values.

FICTION: Your electric car battery won't last for the lifetime of your car ownership. You will have to replace your battery or put up with a huge loss of range after just a few years. Most batteries can't be replaced and therefore the resale value of your car will drop significantly in the years ahead.

FACT: Preconceptions on the life span of modern electric car batteries are often based on the understanding of batteries from 30 or even 50 years ago, not today's modern lithium-ion batteries. Today's electric car batteries will comfortably last for the full average lifespan of a car, easily doing 250,000 miles with less than 10 to 20% battery capacity loss. Electric car batteries today can also have a second life as home and business energy storage. Most new electric cars come with five and even eight-year warranties on the battery's performance. Electric car batteries, in today's modern vehicles, are based on proven technology and millions of miles of real-world usage, you will never replace your car battery.

ARE THEY REALLY CLEANER?

Electric cars need electricity and much of the world's electricity is still produced from fossil fuels across the globe. The argument that electric cars are not really a solution to climate change, and just shift the problem from the gas engine to the power plant has been used for years to resist electric car adoption. I am not saying that this isn't true, 80% of the world's energy is still produced from fossil fuels, even heavily polluting coal-powered generation stations. In the final chapter of the book, I'll explain the great energy transition and how the percentage of 'cleaner' produced, renewable electricity is increasing across the globe. My argument to increase the adoption of electric cars is that although many power stations are still CO_2 producing, they are centralised. Yes, power plants are mostly pumping CO_2 into the atmosphere, however, it is still more efficient to produce energy at a central source than millions of mini power plants in the form of internal combustion engines. The added benefit of electric cars is that the pollution is removed from our city streets. This means we can benefit from clean air, and all the health benefits that brings on a local level where most people live, away from power plants.

The other argument against electric cars is that they produce more CO_2 than internal combustion cars to build. There are multiple sources and methods for calculating the net climate impact of an electric car over its lifetime. Although these models often give different results, in the main, they confirm electric cars are far cleaner, better for our atmosphere and air than internal combustion. There are also a lot of mistruths out there, and deliberate misinformation, often shadily backed or supported by big oil and vested interests trying to discredit the 'cleanliness' of electric cars. Some of the research and models don't perform fair like for like comparisons. For example, they will compare the total cost in terms of carbon produced to build an EV plus the cost in terms of carbon to power its battery over a lifetime then compare this to the gasoline car equivalent. Often these calculations deliberately exclude the CO_2 output from pumping the oil out of the ground, so-called well to wheel, refining it, and then moving it all over the world in diesel sea tankers and trucks to your gas station.

Many media reports have surfaced, often in reputable mainstream newspapers, claiming that electric cars are even worse for the environment than internal combustion. This is not true, and in the regular words of Mr. Trump, fake news, in the true sense. Many of these reports have been spectacularly debunked and proven to be supported by internal combustion vested interests. Once again, I'm not trying to blindly convince you that electric cars are super clean and absolved of all climate impact guilt. Building electric cars do produce a negative climate impact, so does electricity generation. Electric cars must "earn" their green badge of honour. Until all electricity is produced from clean renewable sources, electric cars need to accumulate lots of clean miles before they become net carbon neutral.

To demonstrate the climate impact of an electric car let's start with the manufacturing process. Electric car manufacturing is still nowhere near the scale of traditional gas car making. There are still a lot of inefficiencies in the building of electric cars in terms of economies of scale, labour, and materials. As electric car production ramps up economies of scale will begin to enter the supply chain and production cycle, thereby making the net climate impact of electric cars gradually less and less. For this comparison, I'll forget about these potential future savings and just look at the situation today in 2022.

Electric cars produce more carbon dioxide to build than gasoline cars but then produce fewer lifetime emissions of greenhouse gases than internal combustion, this is a fact. In countries where the energy percentage generated from renewables and nuclear power are higher the emissions level from electric cars can be even lower, up to 70% lower in countries like Sweden and France for example. This is because Sweden has a high proportion of renewable energy and France has a high proportion of nuclear energy. Various mechanisms including life cycle-based sustainability standards have been used to compare the two like International standard, ISO 14040. During the production of electric cars, the battery is the biggest consideration as the chassis and body frame are pretty much like a gas car. Raw materials like cobalt, lithium and rare earth elements are required to build the battery pack. Mining, smelting and preparing these metals produces carbon dioxide. Indeed, it is estimated the current manufacturing process to build an electric car produces 1.5 times more carbon dioxide than the equivalent fossil fuel car. This is 14.6 tons of CO_2 for electric compared to around 9 tons for an ICE car. The battery production alone accounts for at least 40% of the car's life cycle emissions. This is a bad start for electric cars but of course, the total life cycle must be considered.

70% of modern cars today are comprised of plastic, however, it only represents about 10% of the weight. Although plastic has gotten a lot of bad press in recent years, it is a low emitter of carbon, and can even be used as a storage for carbon. Therefore, apart from the battery, much of the rest of the components are built the same way for electric and gas. So, all things being equal what we're comparing here is the carbon cost to produce a battery pack plus the source of electricity to power that pack over its lifetime versus making a gas car, its maintenance, and its burning of fossil fuels over its lifetime.

This is not a straight-line comparison to compare the environmental footprint of either vehicle, as different manufacturing suppliers, techniques and components vary as do sources of electric power generation. The International Council on Clean Transportation (ICCT), the organisation responsible for exposing Dieselgate, state the typical European electric car is at least 30% cleaner than its petrol or diesel equivalent.

This percentage is compared to even the most efficient internal combustion car. This life cycle assessment is based on four stages including 1 - supply chain, 2 - production, 3 - lifetime usage and 4 - recycling. Gas cars win on 1 & 2, but then electric cars win big on 3 and 4, significantly beating gas cars. Electric cars win on criteria number 3 because they are far more efficient at converting energy into movement. Electric cars score a further big win on criteria number 4 because batteries can be repurposed after they're finished in electric cars for home storage and business use (I'll explain all this in chapter 8). Modern electric cars have indeed been designed from the beginning to be disassembled for recycling.

This all means that although electric cars have a bigger carbon debt when they come off the production line, this is usually written off within two years and the vehicle becomes carbon neutral, while its petrol or diesel cousin continues to pollute for its lifetime. The efficiency of internal combustion cars has improved by about a factor of three times since the Ford Model T. Efficiency has peaked with only small marginal gains in the most recent decades. Therefore, gas cars will no longer deliver any efficiency benefit or lower carbon emissions going forward while electric cars will continue to get better as the source of electricity generation shifts towards renewables. In other words, although electric cars are on average 30% cleaner today this percentage is going to grow. Another climate benefit is that with smart charging, electric cars can take advantage of lower peak demand electricity, which is usually cleaner, as it is not produced from typical standby fossil fuel electricity generators.

Battery production will become more efficient, energy sources will become cleaner, and economies of scale will drive the margin between clean electric and dirty internal combustion even wider. An example of this can be seen from the Swedish Environmental Research Institute. In a 2017 report, the Institute calculated that there was an average of 150 to 200 kilos of carbon dioxide produced per kilowatt of battery capacity. Just two years later, in 2019, the Institute revised its estimate down to 46 kilos of carbon dioxide per equivalent kWh produced; this is because the battery factories have been scaled up and are now running at a more efficient capacity. Improved mining techniques and more efficient recycling processes are expected to drive this ratio down

even further in the years ahead.
Before I conclude the argument that electric cars are cleaner, I want to take you back to the misinformation and deliberate deceit we have witnessed from the oil industry and car manufacturers like Dieselgate. Several reports have appeared in mainstream media making outrageous claims that electric cars were worse for the environment. Based on often dubious facts, usually relying on the production of energy from coal power plants and the battery materials, these articles appear to this day and often have questionable backers.

One such article was published in late 2020 claiming electric cars had to be driven at least 50,000 miles before they became greener than their petrol counterparts. The report behind the article, which appeared in mainstream UK press, was commissioned by Honda, Aston Martin and the fossil fuel industry.

Electric cars are not the panacea to avoid a total climate disaster, however, their emergence and eventual dominance are also helping to accelerate more research and transition into renewable sources of energy. Clean sources of energy combined with more efficient clean electric car production are the future.

FICTION: Electric cars are worse for the environment than petrol or diesel cars.

FACT: Electric cars are cleaner by at least 30% over their lifetime than an equivalent internal combustion car. This can grow to over 70% where renewable energy has been used to power the car. The manufacturing of electric cars is currently 1.5 times worse for the environment than an internal combustion car, however, this is recovered over its life cycle through zero emissions. A Transport and Environment study in 2021 suggests that battery-electric cars will use 58% less energy than a gas car over their lifetime and emit 64% less CO_2. Electric cars are consistently becoming cleaner, with improved production and cleaner energy sources, unlike internal combustion cars whose emissions efficiency gains has largely peaked.

ENOUGH ELECTRICITY

"There will not be enough electricity and electricity grids will not be able to cope with millions of new electric cars!" Debunking this myth requires an understanding of how power is generated and delivered. What's generated is used. Excess energy is very difficult, and expensive, to 'store' for when it's needed at peak demand. Storage only represents a tiny fraction of total electricity output at any given time. Instead, power supply operators rely on backup generation capacity to meet peak demands. In other words, most of the time electricity power plant generators are unused or on low output mode, this can vary by the time of day and year greatly.

The two issues to consider are therefore if there is enough generation capacity to meet electric car needs and if the power grids and power lines can cope. On generation capacity the answer is broadly yes, there is plenty of spare capacity. Indeed, spare capacity is increasing in some developed economies due to greater efficiency of power usage with better energy ratings on household appliances, industrial and business equipment. This combined with more renewable and energy storage solutions is reducing peak demand on power plants.

In the UK for example, in 2002 total peak demand was 62 gigawatts, in 2020 peak demand had dropped to 52 gigawatts. EVs are most usually charged at night, at home in at least 60% of cases. Night-time capacity in most countries is more than sufficient to power entire nations driving EVs. It has been estimated that demand would increase by 10% in the UK, for example, if everyone switched to electric cars. Smart charging, whereby charging can be scheduled through the cars or wall charger's software, can be easily set to only charge at low demand, off-peak, and often cheaper electricity times. This would further reduce stress on national power supplies.

Globally the world has 8,000 gigawatts of operational electricity generation capacity. Running all capacity at once can produce almost 70 million gigawatt-hours of electricity. In 2019 the world consumed around 27 million gigawatt-hours of electricity meaning 'spare or peak' capacity is about 40 million gigawatts. It has been estimated that the total power required to run the world's car and transport fleet of over one billion units would be about 28 million gigawatt-hours of electricity. So, in summary, there is plenty of spare generation capacity to switch every car in the world to electric today. However, the current mix of power generation from gas, coal, oil, nuclear and renewables like, wind and solar, would mean limited CO_2 emissions savings if every car ran on electric. The CO_2 emissions drop would only by about 10-15%. What needs to happen is a parallel shift to cleaner sources of energy as we switch over to electric cars. Thankfully this is happening, to varying degrees, around the world and I'll cover this in the final chapter of the book.

Although the worldwide picture seems to imply there is enough power, it varies from nation to nation. During the 2021 Texas power outages, when its electricity grid failed due to unexpected demand because of an unusually harsh cold snap, electric cars were rendered inoperable. Unless EV owners had their storage capacity in their homes, EVs could not be charged. Ironically, had the power to grid services been available en masse, EV owners could have powered their homes from their cars. The issues in Texas were a peak demand limitation as the state never had to deal with such harsh climate conditions for that time of year, arguably a freak weather event caused by climate change. Rolling blackouts also occurred in California in 2020 due to a heatwave as air conditioning power demand stretched power supplies.

The geographic size of the US means there is no single power grid. Regional grids operate independently of each other and cannot share excess power generation. This is common all over the world, in larger countries particularly. Capacity increases are predicted to be necessary for many US states from 2030 onwards. These are not for everyday EV usage but those peak demand events such as the ones in Texas and California.

Overall, the US power system generates excess energy annually, however, unlike in the UK, peak demand is growing. In the year 2000 power plants produced 3.5 trillion kilowatt-hours of energy while in

2020 that number had grown to 4.0 trillion kilowatt-hours.

Today 100 miles of range in an electric car is roughly equivalent to the power used by an average American home in one day, 30 kilowatt-hours, equivalent to around 1 gallon of gasoline or four litres. This is based on an average of 100 miles per gallon equivalent efficiency (MPGe). For example, a Tesla Model 3 standard range plus is the most efficient at 141 MPGe, while a Porsche Taycan Turbo S is the worst at 68MPGe. The U.S. Department of Energy conducted a study that economic growth combined with increased electricity demands, especially from electric cars, could increase US power consumption by almost 40% by 2050. In California, the state predicts that EVs will draw less than 6% of the state's power generation capacity by 2030. In the US individuals cover around 3.1 trillion miles per year, roughly 13,500 miles per person for America's 230 million licensed drivers. The total estimated power requirement if every car in the US was electric today would be around 1.2 trillion kilowatt-hours.

The US could not support an overnight switch to electric cars. That's not going to happen, the switch will be gradual over the next two decades. Even with the exponential shift to EVs, power generation has time to ramp up. This ramp-up needs to see a shift away from fossil fuel power generation to renewables combined with new energy storage solutions for unpredictable peak demand events. This potential additional demand for electricity must be taken into a historical context.

In the US alone power generation capacity was able to increase five-fold from 1960 to the year 2000, according to the U.S Energy Information Administration. This increase was driven by population growth and the widespread use of new appliances, like at home air conditioning. With a 30 year to full electric car transition forecast (although I believe it will be just 20 years) in the US, just a 30-40% increase in capacity is required, well below historical power increase achievements and a modest challenge for the energy industry.

Power generation is one thing, but can the power lines, substations and homes handle the load? Let's take your home first. A standard

220v-240v or 110v (for the US) domestic plug with just a 13/16amp fuse can charge your electric car. This will be slow, likely only adding between 8-15 kilometres (5-10miles) of range per hour. This will drain no more than many household appliances, drawing around 7,200 watts or less. This would be a little more than a home air conditioner unit and less than a water heater for example. For faster charging you will need a home charging unit and small fuse upgrade, typically costing between $500-$1,000, which can fully charge most of today's EVs in 3-4hours.

For high speed and super-fast charging, the grids in most countries can handle the load today. This is because most fast chargers are located along major highway routes where high-capacity power lines are often located. Local grids can handle the increased EV load, however, not at the same time. If everyone arrived home in the evening at roughly the same time and plugged their cars into the local grid, it could fail, especially given that 6 pm - 8 pm is usually peak power demand. Therefore, smart charging technologies are necessary to balance the load and charge cars at off-peak times. Local energy providers could easily regulate this by charging more for peak and less for off-peak usage to encourage EV owners to set charging to start in the middle of the night for example.

Given average driving distances, all night charging is not necessary, 2-3 hours with a standard domestic socket would be enough to meet 90% of daily driving. Of course, this adds a little bit of a hassle factor, however, most new EVs already have this simple charging scheduler built into their software. Using a smart charging solution largely solves grid and power line peak demand capacity. This would mean today's infrastructure doesn't need significant upgrades. None of this considers the increase in home power generation solutions, like solar panels and solar roofs in many parts of the world. These combined with home energy storage batteries, like Tesla's Power Wall, will gradually provide a further cushion to power grids. Indeed, given energy storage and peak demand is the number one issue for energy suppliers, a world full of electric cars could help to balance and reduce peak demand by tapping into unused power in the cars' batteries connected to your home's power supply. Power to grid solutions are already operational in many parts of the world and I'll cover this in more detail later in chapter 10.

FICTION: There is not enough electricity to power electric cars today. The increase in power generation, with the planned switch to electric cars, will not be able to scale up fast enough to produce enough power.

FACT: The world already produces enough excess electricity today to power a world of only electric cars. The power supply today is sufficient to meet the needs of the planned expected shift to electric cars. In some countries, infrastructure upgrades are required, while in others a 20-30% increase in peak capacity is all that is required to power every vehicle in the world. In summary, the world's electricity supply can handle the switch to electric cars with only modest increases in some places compared to historical power generation achievements.

THEY ARE DANGEROUS!

I am sure if you've regularly driven a long journey on the highway, you may have witnessed a burning or burned-out car on the side of the road, typically on sweltering hot days. I have, on a few occasions, thankfully with the passengers standing by the side of the road with safety and fire crews attending to the flaming vehicle. A friend of mine once restored a vintage VW camper van. The rear engine air-cooled machine developed a faulty fuel connection and burst into flames on a major road with himself, his wife and daughter in the van. Thankfully they were able to pull over, and all evacuate safely, only to watch their restored classic go up in smoke.

Petrol is a highly combustible liquid, ultra-refined crude oil designed to explode with minimal ignition effort. Spark plugs ignite a compressed gasoline/air mixture inside the combustion chambers of ICE cars, creating thousands of controlled explosions producing fire, heat, gasses and ultimately movement. Internal combustion cars are not dangerous, however, there is a risk of fire, especially after an accident and we have all heard horror stories of people being trapped in burning vehicles.

Modern cars have a high safety record when it comes to accidental fires, but they do happen. According to the US Fire Administration, there are over 170,000 highway fires per year. Some of these fires are caused by exploding truck tires, construction and agricultural vehicles, but the Fire Department states that one in eight highway fires is a car fire.

In the United States, vehicle fires are responsible for 350 deaths per year and over 1,300 injuries. The risk of fire has reduced over the decades as cars have become safer. Few fires, these days, are caused by design flaws and most are caused by an accident or overheating. There have been some well-publicized mass vehicle fires in recent years. At the height of the Covid pandemic, nearly 4,000 stored rental cars burned in Florida, in 2020, causing over $100 million in damages. In 2018, over 1,000 cars burned in a major car park in Liverpool, England. Both fires were a result of just a single vehicle catching fire and spreading rapidly. Apart from these high-profile cases, car fires happen every single day and generally don't get much publicity. An electric car fire, on the other hand, gets hyper publicity.

Whenever there's been a fire with an electric car, the media, and anti-electric establishment, jump all over the story and decry electric cars as dangerous. I am not going to defend the combustion possibility of lithium-ion battery packs, they can go on fire and explode. Moreover, once lithium-ion batteries catch fire they are extremely difficult to extinguish and burn at extremely high rates. Most electric car fires have occurred following a collision or some extreme piercing of the battery pack. It is extremely rare for an electric car to spontaneously combust, as can happen in a hot gasoline engine. While heat is usually the culprit in a gasoline car fire, faulty battery management systems and connections are usually the culprits in electric car fires.

Managing thermal runaway events in lithium-ion batteries is the primary objective of modern electronic battery management systems. As a lithium-ion battery pack is charged and discharged heat is created; dissipating this heat safely and efficiently is vital for the reliability of electric cars and driver safety. Several high-profile fires have occurred in Tesla vehicles following accidents and piercing of the battery packs. Tesla has claimed its cars experience 5 fires per billion miles travelled versus 55 fires per billion miles travelled in gasoline cars. In the U.S., The National Fire Protection Association and U.S. Federal Highway Administration have investigated numerous electric car fires.

In 2018 the National Highway Traffic Safety Administration, NHTSA, announced that GM was recalling more than 50,000 Chevrolet Bolt EVs because of a potential fire risk. This risk is not limited to US manufacturers, the Audi eTron SUV was recalled following the discovery of potential fire risks due to a wiring harness issue. In 2018, in a Shanghai car park, a video emerged of smoke billowing from a burning Tesla Model S. Another Shanghai parking garage fire, this time involving a Tesla Model 3 occurred in early 2021 drawing more scrutiny on the company's fire safety. Tesla investigated and confirmed the fire was due to a faulty battery module.

Short-circuiting a battery, particularly lithium-ion batteries can cause terminal runaway events known as Joule heating. The chemical reaction that occurs with a short circuit generates more heat than can be dissipated resulting in chemical fires. These can even cause explosions. I have experimented with this myself – I'll explain the outcome in Chapter 8.

In 2019 the London Fire Brigade handled over 50 electric vehicle fires compared to almost 2,000 petrol and diesel fires. This doesn't compare due to the relatively small number of electric cars versus gas. It is conclusive to say, for now at least, the data set for electric cars is infinitesimal compared to gas cars, however, the risk of spontaneous fire is lower overall. Significant work is required by fire departments across the globe to be prepared to handle electric car fires. Lithium-ion battery packs can burn for 24 hours at extremely high rates and cannot be put out with normal techniques used on gasoline fires. Fires have not been the only attention electric cars receive when it comes to safety. There have been claims electric cars can even be dangerous in wet weather due to the possibility of short-circuiting and electrocution. In a near comedic outburst, an Irish politician, known for his outlandish statements, claimed in parliament that "with just one drop of water electric cars are ruined!" There have even been claims about the electromagnetic radiation produced from the discharging of a battery that can cause cancers. Neither of these claims has ever been proven.

The International Commission on Non-Ionizing Radiation Protection (ICNIRP) found that the electromagnetic spectrum output of electric cars is less than 20% of the recommended limit. High speed and electric acceleration is another risk. There is no evidence to suggest that due to the power of electric cars, crashes are more frequent than equivalent sports or high-powered gasoline cars. Indeed, with a host of safety software features electric cars are generally safer than their legacy counterparts.

FICTION: Electric cars are more dangerous than gasoline cars due to the propensity of lithium-ion battery packs to explode and burn. Electric cars are also dangerous due to the risk of short-circuiting in wet weather and can cause ill health effects from electromagnetic radiation.

FACT: Fundamentally electric vehicles are extremely safe and most risk of fire comes following a collision, no different to the risk of fire in a gasoline car collision. The rate of fires in electric cars versus gasoline cars is just 1 in 10. Electric vehicles are safer than ICE due to the lower potential for flammability with lithium-ion batteries versus gasoline. Per mile driven today electric cars are safer than gasoline cars.

THE HYBRID LIE

Hybrid cars are now a dumb idea. They briefly had a place, back in the 1990s and early 2000s, in demonstrating electric technology in a very limited form, but have no place in the world today and should be banned along with gas cars as soon as possible. Why this extreme view? The hybrid car is complex, heavy, and only marginally more efficient than a polluting gas equivalent. With some hybrids you plug them in (PHEVs), others you don't (HEVs). Non-plug-in hybrids are the dumbest of all. The hybrid car came into existence to meet increasingly stringent emission standards, and at the time of their introduction, high gas prices.

From a clean air and climate perspective, buying a hybrid car versus a gas car is like a heavy smoker deciding to quit by going from 20 a day to just 18, the real benefit is minimal. Whenever I take this strong line on hybrid cars, I immediately get a pushback that goes something like "but they are still better than gas, and better than doing nothing at all!" Yes, I agree, they are slightly better than gas when it comes to air pollution and our climate and they proved interesting for a decade or more, however, their time is now at an end. Along with the internal combustion bans planned by many countries, from the mid-2020s onwards, hybrids are also slated to be banned, however typically these bans kick in five years after ICE bans. This is because there is a misbelief that retaining hybrid cars longer is still better for us. It's the opposite. I would argue hybrids should be banned today. The reason I propose such a draconian measure is because hybrid cars are slowing the transition to zero emissions fully electric cars. This is due to two main reasons, number one they are an unnecessary false distraction for consumers, and number two, they are blocking infrastructure which should only be used for fully electric cars.

I'll start with the first reason. Today there is an increasingly broad variety of pure electric cars on the market, capable of catering for almost every need. Range is not an issue for 90% of use cases and range limitations of electric cars are often exaggerated deliberately by internal combustion car makers. The hybrid is therefore used as a placebo, a fake product to make the consumer feel more at ease. By playing on people's fears about the range of electric cars, even if they don't want to continue driving a polluting car, the hybrid appears to have the answer to both. You will never have to worry about running out of range, because you still have an internal combustion engine in a hybrid car, and you can feel good about yourself because you reduced your emissions, right? Wrong.

Research has shown the pollution and climate benefit impact from hybrid cars is negligible. Indeed, it has been suggested that the existence of hybrid cars is encouraging people to buy bigger-engine machines, safe in the knowledge that a tiny proportion of their miles will be covered by electric propulsion. With a plug-in hybrid, PHEV, you charge usually a very small battery that gives a limited range, typically less than 50 kilometres on electric, the rest of the time you're on gasoline. So, 50 clean kilometres is better than nothing at all, right? Well not really, because hybrid cars are using electricity to carry around an internal combustion engine and its fuel. The only slight gain is the small amount of charging from regenerative braking.

The reality is, for 90% of use cases there is nothing stopping somebody from going for a fully electric car and the hybrid option is just an unnecessary distraction that offers no real climate benefit at all. Indeed, it has become quite common for people to state that they will try hybrid first before going full electric due to the irrational fear of range of electric cars. Within weeks of adoption, 97% of people switching to electric cars are used to it and state they would never go back. The hybrid is therefore delaying people's purchasing decision to switch full electric.

The second reason is closer to my own heart and drives me nuts. I'm going to illustrate it with just one of the hundreds of examples of how hybrids **** me off. When not charging at home, I will sometimes take advantage of the charging points in shopping centres, hotels, or city-centre car parks. I had to drive from Luxembourg to Frankfurt, around 250 kilometres away, to attend a meeting, stay overnight, and needed to get back first thing the next morning. The hotel I picked stated on its website that they had spaces for EV charging, so I booked it. I knew when I reached Frankfurt that my battery would be almost empty but was safe in the knowledge that I could charge overnight and be ready to make it back the next day without having to do a supercharging stop. I drove into the hotel, with just 20 kilometres left on my battery, only to find that the only two charging spaces were occupied by two giant SUV diesel hybrids. Sure, you can claim they are perfectly entitled to use these charging spots to charge their cars, but all they're adding is around 50 kilometres maximum of clean range, and don't need to publicly charge, ever. I needed access to make my trip back the next day as quick as possible, I didn't have polluting petrol or diesel engine as a backup.

This occurs all over the world daily and it seems to be getting worse. Whenever I drive into a city car park, or shopping centre complex, I invariably find three out of four available charging points occupied by hybrids, and they are almost always large heavy polluting SUVs. When a gas car blocks an EV charging point they're called 'Gasholes'. There doesn't seem to be an equivalent for a hybrid car driver so I'm going to just call them stupid. By not being able to use charging infrastructure, because it's being blocked unnecessarily by a hybrid, that further slows the uptake and transition to electric cars.

You could argue that any savings made by hybrids have been lost because the engine sizes in hybrids have increased gradually over the past decade. In other words, people are still buying big heavy polluting engines and trying to slightly compensate with a hybrid car. This is made worse by numerous tax incentives, encouraging people to purchase hybrid cars. It is almost ridiculous to see high powered brands, even McLaren, selling hybrid cars. It is greenwashing for them and deceit for the customer. There is virtually no benefit, economical or environmental.

Apart from confusing the customer, delaying a decision to buy an electric car and blocking essential electric car infrastructure, hybrid cars do have a slightly beneficial impact. This is only really the case for pure plug-in hybrids, not hybrids that charge the battery by burning fossil fuel. It drives me crazy, to listen to radio advertisements from Toyota praising their ''self-charging technology." This is like claiming to have invented the perpetual free energy machine. It's 100% marketing. The "self-charging" was 90% created by burning gas from the engine to produce electricity. This marketing hype fell afoul of the Norwegian government in 2020 who instructed Toyota to stop using this false advertising. It's too late for companies like Toyota to try and delay the electric car transition, although their stance may be about to change as rumours, in March 2021, circulated about a possible collaboration between Toyota and Tesla to build a sub-$25,000 electric SUV.
So apart from being the methadone to a heroin junkie, what exactly is

the benefit of a hybrid car? Well, many consumer studies have shown mass confusion when it comes to hybrid electric vehicles, plug-in hybrids, and full battery-electric cars (BEV). I was once asked by a stranger, admiring the Tesla I was driving, if it was a fully electric car or hybrid. The efficiency measure for a hybrid or electric car is best stated in kilowatt-hours (kWh) per 100 miles. A kWh is equivalent to 100 x 100-Watt light bulbs being on for one hour. You pay your electricity bill in kilowatt-hours, so it's the easiest way to measure the efficiency you get from a gas car, hybrid, plug-in hybrid or fully electric car for a simple comparison. The Tesla Model 3 is one of the most efficient EVs, rated at 23 kWh per 100 miles, while the heavier Tesla Model S consumes 30 kWh/100 miles. The part solar powered Lightyear One, which I covered earlier in the book, scores best and is rated at 17 kWh/100 miles while the sporty electric Porsche Taycan is inefficient at 49kWh/100 miles.

So how do hybrids compare to these numbers? The battery capacity of most hybrids ranges from around 27 to 80 kilometres. I'm going to start with non-plug-in hybrid vehicles, HEVs. Generally, at a lower speed, the car engages the electric motor, then at higher speeds, or when the battery is drained, the car activates the normal gasoline engine. In the case of non-plug-in hybrids, the engine not only moves the car but also recharges the battery; this is quite inefficient. It is inefficient for the reasons I outlined earlier; gas cars lose 80% of their output in various ways including heat. So, the car is being charged with very expensive energy, of which 80% was lost. Energy generated from regenerative braking is the primary cost saving in a hybrid car, not the engine battery combo. It gets a little complex in trying to compare the efficiency rating of a hybrid car because they usually have two ratings, MPG from the normal engines, and kilowatt-hours per 100 miles for the electric use. The battery capacity of hybrids is often set at a size that qualifies for tax breaks. In the United States, for example, this is at least 16-kilowatt hours. Batteries this size are significantly smaller, three or four times smaller than in a fully electric vehicle. So, the only real gain from a HEV is regenerative braking. This is taking advantage of "free" kinetic energy, saves brake wear and does reduce carbon emissions, slightly. The batteries are tiny in HEV's, relative to fully electric cars, for example, the earliest Toyota Prius had just a one kWh battery, it has since jumped from 4 to now 8 kWh. The Prius still produces 170 grams per mile of CO_2. So, hybrids offer very limited net climate benefits, although will save you some money on fuel costs with the advantages of regenerative braking. At least HEVs don't block charging infrastructure, unlike PHEVs.

An even more basic version of a HEV is known as a 'mild hybrid'. These are effectively ICE cars with a beefed-up starter motor known as a 'Belt Alternator Starter' (BAS) which recoups some energy lost during braking and feeds it into the battery, offering very limited overall efficiency gains...

PHEVs offer some additional advantages, as on average around 10% of the miles driven can be via the electric motor. The problem is for these limited miles, and relatively little climate saving, charging infrastructure is blocked for pure electric vehicles which are much better in both efficiency and climate impact terms. Looking at a few real-world examples, using 2021 efficiency ratio ratings from the US EPA, I'm going to compare a Tesla Model 3 to an equivalent sized BMW 3 Series Hybrid. Tesla consumes 25-kilowatt hours per 100 miles, and it has a miles per gallon equivalent rating of 134. The BMW 3 Series, 330e PHEV, has miles per gallon equivalent of half the Tesla, at 67. Worse still, when in electric mode it consumes double the kilowatts per 100 miles at 51 kWh. That's the efficiency rating, but what about the climate impact? Per mile driven the Tesla produces zero grams of CO_2, compared to 192 grams of CO_2 for the BMW, while a Prius produces 170 grams of CO_2 per mile. The net benefit is irrelevant in climate terms, and indeed considering hybrids could be slowing the transition to electric vehicles they are potentially having a detrimental climate impact.

A large heavy SUV, like a Range Rover, produces 412 grams of CO_2 per mile, consuming around 16 barrels of oil per year. The equivalent Range Rover Hybrid version produces a much lower 259 grams of CO_2 per mile, burning around 10 barrels of oil per year. 259 versus 412 grams of CO_2 per mile is a significant saving, however, 259 is still a huge climate number and it can be argued that the hybrid label, is disguising the true negative impact. If hybrids didn't exist it is questionable whether such heavy polluting SUVs can continue to be accepted by emissions regulators.

Because hybrids are usually battery and engine technology shoehorned into existing ICE car platforms, weight distribution is very often compromised. On top of this because the electric motor can kick in for just a few seconds, it sometimes delivers uneven torque to the tires, this means hybrid tires wear out far faster; one major global tire brand told me up to two or even three times faster than even electric cars.

Makers of hybrid cars came under scrutiny in 2020 for exaggerating the climate savings of their vehicles. It was found for example the BMW X5, Volvo XC60 and Mitsubishi Outlander emitted between 28% and 89% more CO_2 than advertised, according to independent watchdog tests performed by TransportandEnvironment.org. There have even been suggestions that Europe could be heading for a new Dieselgate, with question marks placed against well-known hybrid brands and the 2020/2021 EU CO_2 emission standards. Tests conducted by the German Fraunhofer Institute for Systems and Innovation Research ISI and the non-profit research organisation International Council on Clean Transportation (ICCT) found that the real-world emissions of PHEV are between two and even four times greater than advertised.

The use of 'self-charging hybrid' terminology in advertising is ridiculous and needs to be banned, as in Norway, while the use of public charging infrastructure for heavily polluting SUV hybrids needs to also be banned and made only available to fully electric vehicles. ICE car manufacturers are hiding behind hybrids to continue to push their fossil fuel drugs. The argument that hybrids play an important transition 'bridge' to new technology is an old, out of date argument. Hybrids did play a part in offering the first semi-alternative to ICE cars, but their time is now at an end, they are 'wolves in sheep's clothing' and a distraction for the electric generation.

FICTION: Hybrid cars are better for the environment, and a good necessary transition to electric cars.

FACT: Hybrids offer just a tiny, inconsequential positive climate impact. Hybrid cars are 1, slowing the transition to full EVs by misleading consumers and 2, blocking public charging infrastructure necessary for full battery-electric cars. Hybrid car technology is now irrelevant.

CHAPTER 7

CHARGING

THE WILD WEST

Yee-haw!!! Here's a sample of the good, the bad and the ugly with EV charging today. AC or DC?, Level 1, 2 or 3?, Charging Speed? Slow, Fast, Superfast, 22Khw, 120kHw, 150kHw or even faster? 800volt charging systems, J1772, CHAdeMO, CCS or DC Combo, 16amps maybe?. Bear with me this is just the start...hundreds of different Charge Point Operators (CPOs - with new ones popping up almost weekly), different connector types, various payment methods, multiple different charging apps, some apps don't show all networks, some show different networks, some apps have some real-time charger status, some don't! Different charging cards, pre-registering, a single charging card for all? – yes but not really. Wildly different electricity prices, your car is mostly blind so no pre-warning if a charger is not working, some different cars have their charging networks, you can't just pay with your credit card, what's OCPP2?... 'error something went wrong'!!!!!! Welcome to the world of charging today – this is what greets a newbie to the new electric car world, welcome to the Wild West!

If you're confused by the above – you should be. Electric cars and charging works are more widespread than you may think and capable of servicing even the longest journeys in Europe and the US…however, this doesn't mean it's easy. When you are sold a new electric car, the guy selling it to you has probably only sold a handful of EVs versus hundreds of ICE cars. The 'product handover' is light, to say the least. You will normally be handed your car key and if you're lucky a charging card, then sent on your merry way – off into the wild west on your new horse alone to find your way! If you buy a Tesla, you are lucky enough to have access to a simple to use, easy to navigate, Supercharging and destination network – which will cater for most of your long-distance needs…but not all.

This is not a Tesla promotion, it's just a fact of current charging life. They have, so far, the easiest to use, simplest, most hassle-free charging solution in place. No buttons, options or payment cards, just plug and play. All the battery range in the world or all the luxury features on new model EVs from the legacy car manufacturers will not matter if the charging user experience hasn't been properly thought-out! I can give you story after story from my own experience, traversing Europe in various EVs, and the experience of a few of UFODRIVE's early first time EV customers plus many others that have told me their charging horrors stories. Range is the number one concern of new potential EV converts. It needn't be. New EVs today have enough range (average 300km) for 90% of your daily needs and long journeys are workable, with a little advanced planning. How much planning is required depends on which EV you buy, the available networks and where you're going.

To encourage people to make the electric leap, charging should be as simple as, or even simpler than, filling your car with gas. Today this is sadly not the case! When you buy a gasoline car the subject of where and how to fill it up never arises, it doesn't need to. You can fill up pretty much anywhere you want. How to find a gas station, if they sell fuel compatible with your car, if you need a special fuel card in advance or if your new car is allowed to use all gas stations and if they are even signposted sounds ludicrous, however, this is how it is today with charging.

EV range, despite all the hype, is largely irrelevant. What's known as 'Range Anxiety' should be renamed 'User Experience Anxiety'. Arriving at a charger that's behind a locked gate, arriving at a charger that's out of order, arriving at a charger that's marked fast – only to find it's been downgraded to slow for some unknown reason. Being routed by your car way off your intended journey only to use a third-party app to get another, better route. Arriving at a charger to find your charging card isn't working or worse, you need to pre-register in advance – sometimes days in advance – before you can use it. And finally, but not an exhaustive list, arriving at a charger to find it's occupied, sometimes even by a gasoline car or a big diesel, climate pointless, SUV hybrid, this all happens daily…why? The reason … lack of user experience design thinking and legacy brought from the ICE days into new electric cars.

We all want simple, hassle-free, time-saving user experiences. We don't want to waste time because of someone else's poor user interface design. Sadly, this is the case with EV's today. We need to stop talking about how big the range of new EVs are, or even if there are enough chargers and focus instead on whether the experience is simple to use. I have lots of personal stories trying to use third party chargers, I'll share just two in the interests of brevity…

My experience number one. I was low on charge, with around 50kms left on the battery. We were going camping the next day (I planned to charge up at home that night in advance), it was a Saturday and I realised we are short a sleeping bag. It was an hour before the camping store's closing time and I lived 40km from the store. Good news though, the well-known European outlet has chargers in the car park, it's a new store, just a year old. My car even showed the chargers on the in-car screen, so no need to use third-party charging apps. I can make it there, charge up for 10-15mins while shopping and make it home, right? Well, the first risk you take with a journey like this is knowing if the chargers are working or occupied. Most 'destination chargers' like these, even, brand new ones, don't have live connectivity to apps or your car so there's a small risk something will go wrong when you get there, a risk worth taking in this case, otherwise I'm sleeping in my coat!

I have a ‘universal’ charging card, I won’t mention the name but it’s fairly well known in Europe...so what can go wrong, off I went. I arrived, with 8km of car battery remaining, at the store and sure enough there stood the shiny new charging posts sunk into fresh white concrete, the lights were on...but was anybody home? I reversed up, took the big blue charging cable out from the trunk and connected the car and charger then went to the touch screen. It’s in Flemish!, I was in the French-speaking Wallonia region of Belgium, however, for some reason, the screen was in Flemish (for non-Europeans Flemish is a mix of Dutch and French, from northern Belgium). Nothing was happening (as is normal before you swipe your charge card). It was easy enough to figure out to press number 1 for the number 1 charging plug I chose. I pressed and swiped my ‘universal’ charging card. Nothing. No juice! I tried a few times then pressed a few buttons randomly but still no juice, it was lightly raining by the way and unlike most gas pumps, EV charging points are rarely covered, again for some unknown reason.

It’s now 15 minutes to store closing time, so I temporarily abandoned it and went into the store to solve it later. Now getting dark and raining more heavily, I went back to solve the charging problem. No joy... and I couldn’t unlock the charging cable from the charger no matter what I tried (this is normal until you close a charging session to avoid someone stealing your cable). The only option was to scan the QR code on the charger and try to figure it out. The website that popped up, on my phone, was also in Flemish and French, I speak neither. Last resort, call the website number...so I did. It took around 15mins before I finally got an answer and luckily the chap on the other end spoke English. He explained that I needed to pre-register to get one of their charging cards in advance (which needs a Belgian address which I don’t have) and it takes a few days. He also explained that he wasn’t sure if this network was ‘on’ the ‘universal’ charging card providers service yet. There was nothing he could do, so there I was, in the dark, wet, almost juiceless and marooned. The only way to remove my cable was to perform an unusual ‘hack’ the guy explained over the phone. It involved pushing and pulling the charging cable in a certain sequence to cause an unlock (I’m not going to tell you exactly how as I promised him I wouldn’t and it’s only for emergencies). I managed to crawl to a nearby guesthouse and the manager very kindly let me plug my cable in the window to a standard 220v socket – where I stayed almost two hours to get enough charge to make it home. Nightmare experience, three hours lost, brand new hardware which was unusable – this happens daily in the Wild West.

My experience number two. I was bringing a brand new, €100k, well known German brand electric SUV from Amsterdam to Luxembourg, a distance of around 400kms. The car's route planner picked a 'fast charger' roughly two-thirds of the way in Eindhoven. I arrived at the charger, with about 30km left on the battery, to find one brand new fast charger installed, however, not turned on, and a second supercharger that was still a big electrical cable sticking out of the ground. Getting out of this one involved a tow truck!

Hopefully, these sample stories paint the picture, it's not about range or the number of chargers it's about CONNECTIVITY coupled with the certainty of a good user experience.

With Tesla, you arrive at a supercharging station then plug and play. There are no touch screens or options on the charger, you know in advance if they are working or occupied and the experience is consistently simple. Other, even faster, supercharging networks are popping up all over Europe and the US, and in the main, they are pretty good but the new car interfaces are not. VW chief Herbert Deiss publicly lashed out at this in mid-2021, he even seem surprised by the issue! Different car makers are doing different deals with charge point operators, some cars show different chargers on your route and pricing for electricity prices varies wildly – usually two or three times what you pay for home electricity, but, sometimes 5 to even 10 times your home electricity cost. To highlight this, try playing around with some charging maps online or charging apps – you will get very different results. I'll pick one, abetterrouteplanner.com. If you choose a Tesla and a similar equivalent EV and pick a fairly common highway route, say Chicago to Washington DC or Paris to Amsterdam or Berlin, you will in most cases see the Tesla arrive earlier, sometimes by hours. This is not because it's got a better range, is faster or even has the fastest charging network, it's because the UX is simple and universal. Competitor cars will route you to multiple COPs and you get there later – in most cases.

EASE OF USE (and why Tesla is so far ahead)

Let's look at why this is so important and how it can be done, not perfectly but by far the simplest so far. Tesla's approach to building cars was to start with the battery pack, then build the user interface and then build the car chassis around that. With none of the legacy carmakers' heritage, Tesla truly started from the ground up. For example, Mercedes released its EQ electric series, which was mostly the legacy chassis and designs from the ICE past, e.g., the Electric EQC is strikingly like the GLC because it is the GLC with batteries. Only with the 2021 release of its new flagship EQS did Mercedes start from the ground up.

From the start Tesla didn't just reimagine a new form of a gasoline-powered car, they reimagined an entirely new way of interacting with mobility. Anyone, within just a few minutes, can figure out how to use a Tesla's main console. It's a simple point and touch, smartphone-like interface. If you use the Google Maps navigation interface to find a destination outside of your car's battery range, the car will point you to the necessary Tesla supercharging stop. This is a bit of a limitation as Tesla currently only show you their supercharging network, and don't make it easy to find other third-party supercharging networks, however, this will change in the future. The car has an easy-to-use energy screen, that will display, based on how you're driving, when you will arrive at your charging station and how much range you will have left. If you drive too fast the car will warn you to lower your speed as you may not make it to the charging station. When you get to the charging station you simply reverse up to the white and red charge point, unplug the cable just like a gas pump, and plug it into your car. The car does the rest, showing you how fast it is charging and when it will be complete. You don't need to think about anything else.

I can't emphasize enough just how this simple ease of use is critical to encourage people to adopt electric cars. Tesla got this right and with constant over the air updates, it's getting even better. You can even see in advance how many of the charging stalls are currently occupied, out of order and available. The charger speaks to the car, the car speaks to the driver, and you don't need any pre-knowledge. This is not the case for almost every other new electric car and almost every other new third-party supercharging station. Indeed, many of the manufacturers, particularly the German ones, are over-engineering the user experience and adding lots of, unnecessary options, which just complicates things for the user. After all few of us use every single option on our microwave oven, TV, smartphone, or laptop for example. Tesla has kept it simple, tailoring their experience to 80 plus per cent of use cases, and haven't tried to build options for every possible eventuality.

Other manufacturers have unfortunately allowed their brand new EV's to be contaminated with the legacy of their internal combustion history. I have driven lots of the new electric car models, and yes you may say I'm just used to the Tesla, but trust me, intuitive these are not in most cases. As you can see from the stories earlier what's worse is sometimes the navigation to third party charging networks is very unpredictable. When you get to some of these other chargers, they often have nice flat screens with lots of options and buttons. Sometimes you're asked which type of connector you would like to use, like a CCS combo or DC. Sometimes you are asked how many amps you want to charge at, sometimes you connect your cable to the car then find the charging card won't work.

A friend of mine finally decided to take the plunge in 2020 and buy an electric car. He bought a very expensive luxury new electric car from a major German brand. After about two months he called me up and said to me "how the hell do your customers avoid having a nightmare in these electric cars?" I asked him to explain, even though I knew already what he was going to say. He had taken his first long trip about 500km/310 miles with his kids. He said he had a nightmare charging, the first supercharging station he went to didn't accept the charge card he was given when he bought the car. The second charging station he went to was in a hotel complex, the charger was out of order, however, the car's navigation system didn't know that. He ended up calling into a roadside cafe and plugging the slow cable into a standard 220 Volt plug and waiting for a couple of hours before he could resume his journey to find a fast-charging station – sound familiar?

He was utterly dismissive of electric cars doing long journeys, he went as far as to tell me he thought they were a fraud. He even contacted them and asked them to take the car back. By the way, when they sold him his brand-new electric car, they gave him practically no advice on how to use it; these sales guys were gas car sales guys. His experience is not unique, these stories can be found on multiple consumer websites and EV forums. Indeed, a high-profile example with the new Porsche Taycan was written about in a major UK newspaper in 2020 after a couple spent hours trying to find working chargers. These stories only perpetuate further the myth that electric cars don't work for long journeys and give false credibility to internal combustion diehards. The unfortunate fact is, these electric journey disasters are nothing to do with how many charges there are, or the range of the cars but is everything about simple easy to use consumer interfaces. I honestly don't get it, legendary OEMs have had a decade to replicate Tesla's success, yet they come up with these unusable user interfaces. I never saw the Blackberry age coming. I got one to trial in 1998 for a bank I worked for and couldn't see the need for email on demand outside the office. It wasn't the way we communicated then. How wrong was I. Years later I did see, however, the inevitable success of the iPhone and the death of Blackberry and Nokia (effectively) in less than five years. It was simpler, and easier to use the platform and offered an endless possibility of opportunities for mobile software applications. As the iPhone began to become the smartphone standard, legacy handset makers still clung to their designs and operating systems. They continued in the misguided belief that their loyal customer bases wanted keyboards or twenty touchpoints instead of two or three. When introducing new technology, simplicity and ease of use must rule. Always.

The existing car brands are Nokia and Blackberry today. They have been asleep at the wheel. They initially laughed at, then became curious and are now scrambling to replicate as they watch Tesla's meteoric rise. This OEM scramble to adapt is analogous to Blackberry putting a touchscreen together with physical keyboards on their iPhone killer devices. Every attempt failed. Repeatedly the words 'Tesla Killer' have been written and spoken at each launch by an OEM of their electric offering. Lipstick on a pig. That's the closest analogy that comes to mind when I see the attempts by big-name manufacturers to shoehorn electric drive trains into existing chassis platforms. I'm not saying these new cars are inferior to Tesla, on the contrary, they are in most cases better built and have much higher quality standards. But never mind just the platforms, they are trying to shoehorn electric into their DNA. It doesn't work.

The only way to get the user experience right is to completely drop every piece of thinking from the ICE Age. Imagine if today, with your gas car, if it needed a multitude of different fuel cards, for lots of different gas stations, where you had lots of different choices, types of fuel, speed of refilling options, choice of fuel chemical composition, all combined with some fuel stations being randomly closed, not working or out of order without notice... you would go crazy. The user experience is vital. Like an iPhone, the new product must be easier to use, understand and enjoy than the incumbent. There is still a lot of work before all of this becomes standardized, and it needs to become standardized and reliable for everyone to use EVs. The good news is there are lots of solutions in the pipeline to make this simpler, and it is likely that Tesla and others will share access to their supercharging networks as time progresses. Tesla is not just way ahead of the competition because of its simple user experience, but also because of the mountains of data, they have been collecting for over a decade. Tesla has been recording and analyzing customer journeys and using this data to develop its interface even further. The other manufacturers are way behind on this.

To further demonstrate this point, and again I'm not trying to blindly sing the praises of Tesla, let's take a two-car challenge. Let's imagine you had two electric cars, both with the same battery capacity and range. Let's imagine both cars performed the same on range, HOWEVER, one was a Tesla, and one was one of the new German, US or Korean

EVs. Let's then say you then enforced a fixed speed limit of, say 120 km/h or 75mph and the driver could only stop when he needed to charge. Now let's put the two cars at the same starting point, with the same traffic conditions, and on the same highway route, say Paris to Berlin – 1,054 Kilometers, 655 miles, about the same as driving from Chicago to Washington DC. Assuming both cars drive around the same average speed, I would guarantee that the Tesla will arrive in Berlin before the other car. Why? The answer is because of the efficiency of the navigation, Tesla supercharging network, and ease of use. The competitor will have to use several different charging networks, and there is no guarantee of connectivity and speed.

I haven't just come up with this hypothetical situation myself. Online EV enthusiasts and bloggers have done such challenges and proved my hypothesis. Using abetterrouteplanner.com, you can see a Tesla Model S can complete this journey in 10 hours and 21 minutes with four Tesla supercharging stops. A Mercedes EQC, for example, will take 12 hours and 15 minutes, with six fast-charging stops. These fast-charging stops will be with a mix of charging providers, from Ionity to BP and will require a charging card to activate them. By the way, the same cars travelling from Chicago to Washington will take just three supercharging stops and 12 hours and 28 minutes for the Tesla but will be 14 hours and 25 minutes with five charging stops for the EQC. Google Maps show this drive to be 10 hours and 38 minutes in a gas car, before you add rest stops. This means that the cost to you in terms of time, to do this long journey in the Tesla, is around an hour, or 10% of your journey time. That will be almost three hours cost to you in the non-Tesla. This simple test demonstrates the work still to be done in making charging simple and just why, for now at least, Tesla is so far ahead in the EV experience. Only start again thinking and making the experience even simpler to use than a gas pump will help convert people to electric cars eon masse. The OEMs will eventually all standardize their navigation and charging systems, however, for now, drivers need to be wary. So how does it all work today and what's happening in the world of charging? I'm going to start with where we charge today and will in the future.

WHERE TO CHARGE

For this next section of the book, please try and forget everything you know and are used to with your gasoline car, gas stations and gas pumps. Try to imagine Electric is the way it has always been. I'm going to explain where to charge, how to charge, what's happening now and what's happening next. Let's start first with where you can charge today. Unlike gasoline cars, where you can only refill at a gas station, with electric, 'filling up' can be done at multiple locations. While filling up with gas only takes minutes, charging can take hours, but does that matter?

We can't directly compare the process to refueling at gas stations because as I explained earlier the use case is completely different. We don't have gas pumps in our homes and usually don't have gas pumps in shopping centers, or hardware stores, or city center parking lots. It would be quite handy if there were lots of gas pumps wherever you went, you'd probably only fill up little amounts when you need it, but when I drove gasoline cars, I always filled up to the very top, and always let my car run to practically empty, because I wanted to save time going to gas stations. It's kind of the same for electric. So where will you be charging your electric car? In the most likely order:

1. At home
2. At work
3. On-street / public charging
4. Destination chargers
5. Superchargers

For most of us, charging will take place at home. If you have a garage or driveway, you can easily just use a standard 220v-240v (110v in the US) outlet, which will add around 100km to 150km (90 miles) of range overnight while you sleep. This is known as 'Level One' charging with simple electric vehicle supply equipment (EVSE). This is what I do. If you want to have a faster charge rate, then you will need to spend up to $1,000 to get a 22kW charger installed that will charge most electric cars from 0 to full in about three hours, this is known as Level 2. When it comes to funding the installation of domestic home chargers, many governments offer subsidies or grants. On top of this many OEMs include the cost of a home charger as part of the deal when you buy a new car.

According to the US Office of Energy Efficiency and Renewable Energy, 84% of current EV drivers charge at home. What happens when you don't have a garage, a driveway or parking space outside your home? It's not easy to exactly quantify how many people will not be able to charge at home, the data is not precise. One way to figure this out is to look at how many people live in single-family ground level dwellings. Starting with the United States, on average 70% of the population lives in a single-family home. This would imply that at least 30% of the population cannot charge at home. If you assume the distribution of car ownership is even amongst homes in the United States and all gasoline cars were replaced with electric, this will mean that over 80 million cars could not charge at home. This will vary from state to state, for example in New York almost 25% of the population live in apartments, while in West Virginia just 5% live in apartments.

In Europe, the picture is even worse, where just 57% of the population live in houses and 42% live in apartments. This means almost 120 million cars in Europe will not be able to charge at home (assuming of course there is no basement charging or equivalent on street). So where will they charge? Let's start with work.

I had to attend a meeting at a brand-new, 20 floor headquarter building for a major European bank. When I drove into the underground parking lot, I couldn't believe what I saw; at least 50 car park spaces all with brand-new electric car charging points. Most of them seemed to be turned off and to be fair most of the spaces were occupied by gasoline cars. When I asked about the chargers they said were planning for the inevitable electric future. This is the kind of planning which should be mandatory for any large new buildings, not just office blocks, but shopping centers, sports stadiums, parking lots etc. This is sadly not the case today.

For example, in a brand-new shopping complex in the center of Luxembourg, there are around 700 car park spaces, but I counted just six charging spots (most of which are usually occupied by plug-in hybrids!) All governments should make it mandatory in all new building codes that car park spaces have most spaces fitted with charging. Following this all-existing building complexes, especially apartment blocks should have charging retrofitted. This doesn't need to be level two chargers, which could add significant costs, even standard domestic wall sockets would be enough (because most people are going to spend 5 to 10 hours at work). Who should pay for all this electricity? That's another question. It could be part of an employee's package, it could be part of the car park fee in the car park, or could be even supplemented by governments, instead of subsidizing the fossil fuel industry.

So, if you don't have access to home charging, and for now at least you don't have access at work where else can you charge? This is where it starts to get messy; let's look at public on-street chargers.

Many major cities have installed on-street public charging. Some of these are supplied by the city, and some by private operators. The same problem exists as described earlier, ease-of-use, whereby multiple charge cards and a lack of visibility of the charger status is the current state of play. Also, many of these on-street charging points are often occupied, by gas cars or 'hogged' all day by a single EV user who left their car plugged in for hours. Some operators enforce penalties and surcharges if the car has been parked charging for too long. Some cities give free parking at EV charging spots, while others charge for parking rates per day, making it prohibitively costly to leave your car there for 8 plus hours. In any event, free car park spaces are not sustainable in cities as the number of EVs begins to grow.

So, what's the solution? There have been many innovative trials of simple technology for charging. One obvious example is to utilize lampposts. In London, lamppost charging was introduced as a trial several years ago. While this is slow. Level 1, charging, does provide a solution, although cables can look unsightly and dangerous lying about on the payment. Pop-up charging points and curbside chargers called 'armadillos' have also been introduced. These chargers are typically dumb, as in not smart or connected so you can't see the status in advance. Lamppost charging in car parks and on city streets is a relatively simple, cost-effective way of providing charging to the masses. As I write this section of the book, I am peering out the window of a Dublin apartment where I have no access to charge a car, however, I can see eight on-street visitor car park spots each with its own lamppost that would be perfect for chargers. It's not just lampposts, many cities are finding new uses for old infrastructure. For example, in Germany, Deutsche Telekom plans to convert 12,000 on-street switch cabinets into charging stations.

So, if not at home, a lamppost, on-street or at work where else can you or could you charge? Known as 'destination chargers', these level two chargers are often found at hotels, restaurants, tourist and leisure locations. The problem is they can be difficult to find as they often don't appear on charge maps because they are usually private. Sometimes when I'm booking a multi-day trip, I will pick a hotel that has charging facilities on site. This takes me a little bit of work and research. I often book a hotel on one of the well-known travel sites like booking.com. Here I can filter by almost every possible choice of hotel facility like breakfast, bedroom type, self-catering, shopping street access, even if the room has a washing machine, however, you can't filter if the hotel has chargers. I wrote to them and asked but never got a reply.

Destination chargers are very handy, they just need to be easier to find. There is a lot more work to be done in making sure everybody can switch to an EV and can access charging at or close to where they live. It will take years before this is in place, but good progress is being made. Like all new technologies, multiple standards will eventually standardize. It is likely in the future that on-street, parking and destination chargers will be so widespread that you never give range even a second thought. The final piece of the charging puzzle is perhaps the most important of all, as it eliminates the much-feared range anxiety. Fast charging or Supercharging.

SUPERCHARGING

Level three charging, or fast charging, is at the top of the food chain. Fast charging speed varies a lot and can be limited by the electrical architecture of your EV. Typically charging speeds of at least 50kw, 120kw and above are possible with supercharging, which can charge an average EV in an hour or less. Achieving 80% charge from zero is possible in 30 mins. Supercharging is perhaps the most important part of the transition to a world of EVs. This also brings the traditional gas station's future into play.

Tesla began rolling out its supercharging network as soon as its first production cars hit the road. Elon Musk and the team knew that to convince people to buy, they had to have the infrastructure to support long-distance travel. This was a bold and necessary move, costing billions in capital. In the early years, Tesla sold its cars, mainly the Model S and Model X, with free lifetime supercharging. With its early lead, Tesla now has the largest fast-charging network in the world. In over 2,500 stations worldwide Tesla drivers can access almost 25,000 individual superchargers, two-thirds of them in North America with the remaining spread evenly between Europe and the Asia Pacific region. Tesla's supercharger works off a 480v direct current technology. This can deliver from 70 kW up to 250 kW of direct power. The power supply is not constant. This is because, to protect the battery, and prioritize who needs the most power first, a Tesla will charge from 0 to 80% in between 30 and 45 minutes. After an 80% charge level is reached it begins to slow down. The design is intended to give the driver enough range to make it to the next supercharger station rather than fully charge the car. Superchargers are also connected in an array which means that it's often first come, first serve in terms of maximum juice. Indeed, a sort of 'gentleman's agreement', has formed whereby very often drivers will knock on the windows of other drivers and ask them if they are almost finished if they need to charge quickly.

Superchargers are not intended to be used regularly. They are designed for, and usually located on, major highway routes. If you use superchargers continuously then, in the case of Tesla, the car will alert you that this is not healthy for the long-term life of the battery. The same goes for constant 100% charging. (I'll explain this in the next chapter). There are enough Tesla supercharger stations now installed in Europe and the USA to embark on almost any major highway long-distance journey.

It's not just about Tesla, other car manufacturers, service stations, public and private operators are now rolling out competitor supercharger networks. In Europe, IONITY for example is a joint venture between BMW, Ford, Hyundai, Mercedes and VW. The IONITY network is open to anyone to use, including Tesla. The 350kW power network uses the now-standard European CCS connector. IONITY's goal is to have a fast charging stop at least every hundred miles along all major highways in Europe. Already over 300 are operational and the network is rapidly growing. Other operators like ULTRA-E are also spreading fast chargers across major corridors in Europe. In the US companies like EVgo and Electrify America are rolling out fast chargers. The current gap between fast chargers along major highways in the US is greater than the equivalent in Europe. America is likely to catch up the pace with Europe as the transition to electric accelerates.

As of 2022, Tesla still has a massive lead in the number of fast chargers in the US, more than other providers such as ChargePoint, Blink and Electrify America combined. In terms of locations though, others like EVgo are just behind Tesla's 900 stations. The biggest issue with supercharging is that some OEMs, like Tesla, are exclusive to Tesla owners while other networks are open to all. It seems ludicrous that even now, new OEMs will start building new fast charger networks instead of sharing. For example, Rivian announced in early 2021 that it now plans to roll out a network of 10,000 charging stations across the US and Canada by the end of 2023. Echoing Tesla's strategy of private Supercharging and public destination charges, Rivian also plans to install 3,500 DC fast chargers' "waypoint" Level 2 chargers. These will use the J1772 plug, now the North American standard versus Europe's CCS standard. What was probably a necessary strategy at the beginning for Tesla to attract customers by building its supercharging network, is not a defensible strategy anymore. With so many new providers coming on the market for fast charging, it's only a matter of time before Tesla and others are outnumbered. Therefore, the number of supercharging stations won't be a competitive advantage in the long run, and OEMs should pool together now to share all this infrastructure rather keep it exclusive for any one brand.

There's been a lot of hype around the concept of ultrafast charging. These systems can operate up to 920v and 350 kW or above. Not all car architectures can take these speeds yet, which aim to 80% charge an EV in 15 minutes or less. The Porsche Taycan, for example, has been fitted out with a 900v system and Hyundai is proposing 800v. The obsession with getting supercharging to be as fast as filling gas is the wrong focus. If you fully charge at home before you head out on a long journey, your first stop won't be for at least two hours, by which time you have done around 300 km (almost 200 miles) and will likely need to take at least a 10–15-minute break. If you make that break twenty minutes to thirty-five minutes, for say a coffee, you will have another two hundred plus kilometers of range added to your car with today's fast chargers. Faster charging comes at the expense of the battery. Many start-ups are working on technology to potentially charge a car from 0 to 80% in five minutes or less, I'll discuss these in the next chapter. For now, I don't think ultrafast charging should be the focus. For most long journeys, you're only going to charge once, on a rest break of say 25 minutes, so it's perfectly workable. More stations are more important than more speed.

Supercharging is an essential part of the electric revolution. Without it, long-distance journeys wouldn't be workable and thankfully it is well advanced in providing a fully workable infrastructure. The focus needs to be on the solutions for those who cannot charge at home. There is a misconception that thousands of supercharging stations are required before electric cars are workable, this is not the case. There is also a misconception that unless you can charge your electric car as quickly as filling up with gas, electric cars won't be practical. The focus, and much of the media attention today, is on developing extreme rates of charging. This is not necessary for 90% of use cases.

THE FUTURE OF THE GAS STATION

Pharmacies began selling gasoline as a side business in the late 1800s. On Bertha Benz's famous first automobile trip she used the first filling station in the world at a pharmacy in Wiesloch Germany back in 1888. Tesla has installed most of its supercharging stations at bespoke sites, often at hotel car parks and retail complexes next to highways. Few Tesla superchargers are in normal highway gas stations. When the whole world is electric what is the future for traditional gas stations? Some argue that there will be a massive contraction in the number of gas stations. Others argue that it will be a structural shift, many smaller city gas stations disappear, while mega new electric stations will grow. The modern highway gas stop has come a long way from being just somewhere to fill up your car. Major highway stations now come with food courts, restaurants, children's play areas and even cinemas.

On a 500-mile/700 km drive, at least one stop is necessary for the obligatory toilet break. A cup of coffee and maybe something to eat and refresh yourself will take you at least 15 to 20 minutes. Major gas station owners see the electric revolution coming. They see the threat to their traditional model, however, also the opportunities to build and supply new services to EV drivers. Food courts and restaurants are obvious ones, but also retail stores, rest areas and meeting rooms. One of the big problems, however, is that many gas stations are in the wrong location. For example, big gas stations just outside cities are probably not going to be necessary when all cars are electric, as most people will have charged at home before they head out. In contrast, highway gas stations will need to be bigger. There have been lots of announcements and partnerships formed to transform the gas station experience. In early 2021, Shell unveiled its plan to roll out over half a million charging points by 2025. With almost 50,000 gas stations worldwide, Shell intends to invest $6 billion to install charging in almost all its stations. Fellow oil giant BP acquired the UK's largest electric charging company in 2018, ChargePoint, and plans to roll out charging across its entire service station network.

Retrofitting old gas stations for EVs is not the only news in town. Purpose-built highway stations for only EVs have also emerged. In the UK, GridServe opened its electric Forecourt in late 2020. The EV-only facility has multiple charging stalls, covered by a forecourt roof with end-to-end solar panels. Facilities such as meeting rooms, restaurants and shopping are on site. The site in the south of England is one of 100 planned. Similar electric-only stations are also now operational in Germany. These purpose-built facilities will likely serve as the future template for all highway stops. In the meantime, in Germany, regulation mandating that gas stations install electric charging is planned. Today EV charging at gas stations is the poor cousin to the gas pump. Often the charging points are usually hidden out of the way, banished to a dark corner. These are usually uncovered and exposed to the elements. I don't know why this is, I have enquired and been given answers ranging from access to the power supply and even safety away from gas pumps! I stopped to charge at a huge highway service station in Germany, the chargers were several minutes' walk from the facilities, and it was raining. It was like I was being punished for driving EV while my polluting counterparts were dry under the forecourt canopy.

The future of the gas station, or rather the future purpose of the ground formerly occupied by a gas station, will be electric. Some will convert existing stations directly to electric, as with the first-ever gas station to be converted to electric in Maryland USA back in 2019, while others will disappear because they are simply now in the wrong location. It is likely that due to home charging, the number of gas stations will significantly decline while the size of gas stations, along major highway routes, will increase. To entice customers some retail operators even offer free charging for their parked customers, IKEA is an example. The gas station will disappear by mid-century, and a new era of highway mega charging sites will reign.

HOW MANY DO WE NEED?

In the previous chapter, I addressed the question of is there enough power. The question of how many chargers is needed is trickier to answer. On top of this, who is going to pay for all this infrastructure and how much will it cost? To answer this question precisely we need to calculate how many people could charge at home. Even if they charge at home, what's the percentage that will need to use fast highway charging for long journeys? On major highway routes, at peak times of the year, will charging stations be jam-packed with queues of waiting EV owners?

With between 30% and 50% of people not having access to install a home charger, public charging infrastructure is vital. There have been many consumer reports stating that EV owners may not have considered buying an electric car if they couldn't charge it at home. Therefore destination, public, on-street and workplace charging needs to be rolled out on a massive scale. It has been estimated that between $100 to $200 billion must be invested globally by 2030 to satisfy EV demand. Charging availability in commercial locations needs to be not just common but mandatory. Anybody operating a public, or even private business car park may be mandated to install a percentage of chargers. It is foreseen that a significant proportion of EV drivers will charge at work, and in the leading EV markets like China, the EU and USA this is likely to be the key focus area in the years ahead. As I explained already most highways are well serviced by supercharging stations, the issue now is more if there are enough spaces and power per station.

I have rarely visited a supercharging stop with more than 10 or 15 individual chargers; this will need to grow by a factor of 10 at least to meet anticipated demand. While most local grids can take level one and level two charging, supercharging is a different story. Highways are usually well serviced by high-capacity power lines but off highways will be a challenge for new fast-charging stations. Another constraint will be the existing car parks' power supply. As I have come to discover (when planning our electric rental locations) many car parks can be 30 plus years old. They were not designed for EV charging, usually, just enough power for elevators and lighting as no heating is required. This is a challenge as I've seen when trying to install 10 or more chargers in an old airport or city center lot. If you take a typical 500 space major downtown parking complex and assume at least 1/4 of the spaces should be equipped with charging, that would be 125 x 22 kilowatts for level 2 chargers. While national power supplies and grids could cope with EVs, if everybody plugged into these 125 spaces all at once there will be over 2000 kilowatts of energy drawn at once. This is not an insignificant amount. So, if integrating EV charging into existing building infrastructure could be problematic and costly, where does this leave us in terms of meeting the demand for those who can't charge at home? The answer is software-controlled smart charging. If we take the hypothetical car park example again and 125 EVs are all plugged in at once, then they cannot all be treated equally. Given that 50 kilometers or less per day is the average mileage driven, there would have to be automatic limits on how much can be drawn per car. For example, with smart charging technology, an EV that plugs into this car park that already has 200 kilometers of range on the battery would just get a small ration of trickle energy / slow charging. Conversely, an EV that arrives in our hypothetical car park and is almost empty would get the full 22 kilowatts of juice up to say 100 kilometers of range. This automatic load balancing is a relatively straightforward solution. Using smart charging, the needs of the many can outweigh the needs of the few. In high-density cities where apartment dwelling represents a high percentage, for example, New York, the increasing use of shared mobility helps to offset the demand for charging stations. People living in high-density cities like New York or Hong Kong typically don't own a car and avail of shared, rental and micro-mobility.

Current estimates suggest that up to 30 million charge points are needed by 2025 in the USA, China, EU and the UK alone. These would be level one and two chargers AC and DC fast Chargers (DCFCs). These charge points will be a combination of destination, home, work, public-private and on street. The significant capital outlay required for all this infrastructure could be a barrier for poorer countries to make the shift to electric. Apart from the individual home, the cost of this infrastructure will likely have to be supplemented by national governments to encourage car park operators and commercial premises.

In a boost to expanding charging opportunities, President Biden's $2 trillion infrastructure bill promises to install at least half a million new charge points across the USA by 2030. Why should the taxpayer pay for the installation? The reason is, unlike gas stations, charging needs to be widespread in thousands of locations. Gas stations can recoup the capital cost of their investment by selling fuel and other goods and services, if commercial premises like car parks had to recoup the cost of a massive infrastructure investment car parking prices would rise significantly. The same is true for workplaces, where a burden will be placed on companies to install and pay for worker charging spaces. If these investments had to be recouped by levying a percentage on the electricity cost, it could make the switch to electric less attractive. So, government intervention, just like Biden's infrastructure plan, is necessary. Charging infrastructure should be considered in the same manner as power lines, rail transport, roads, and waterways i.e., as essential national infrastructure.

COST

Apart from charger availability, another issue today is vastly differing prices to charge your car. The cheapest form of charging is usually at home. For example, in the United States the average home electricity cost is around 12.6 cents/kWh, this means less than $10 to fully charge an average EV. In the EU the average cost of electricity varies greatly between countries. The average for the 27-nation bloc is 22.03 cents per kWh, however, this can be as low as just €.09 in Bulgaria for example. The most expensive place to charge an electric car from your home is Germany at an average of 30 cents per kWh. This means in Germany it will cost around €6.27 for every 100 kilometers of charging, while in Bulgaria this cost is just €2.

However, out on the street or out on the road these costs vary massively. Taking the UK as an example, the cost of home electricity is on average 14.37p per kWh, meaning it would cost around £10 ($14) to charge a typical 70kw battery EV. However, if this car was charged in the UK using a Tesla supercharger this cost will be almost double at £19.60 ($27). This is based on the average supercharging cost in the UK of 28p per kWh. This is a similar case in the USA, the average supercharging cost is around $0.25 per kilowatt versus $.126 for home charging. If gasoline prices varied this much there would be a revolution!

In the United States, the cost of DC Fast Charging (DCFC) charging can vary from as low as $0.10 per kWh to as much as $1.00. For many providers like Tesla, EVgo and Blink, the average DCFC cost is $0.35 per kWh. This means the most common on-street charging will cost as much as three times more than charging at home. The most expensive place to charge your car at home would be Hawaii at over $0.30 per kWh, while the cheapest would-be Oklahoma at just $0.09 per kWh. Another example, this time in Ireland, the state electricity supplier charges 0.33 cents per kilowatt-hour to use a mid-level 50kW fast charger versus the home average electricity cost of €0.24, therefore 50% more expensive away from your home. While the use of on-street and public charging will be the exception for most it will still be the only option for many. Outside of work, the 30 to 50% of people who will not be able to charge at home are going to be subject to costs two or three times more than the people that can charge at home. This is again an example where governments can step in to help level the playing field. For those who wish to switch to zero-emissions mobility, but don't want to get hit by the extra costs of charging away from their home, governments could help to incentivize charging operators for people who don't have access to home charging. A more daring solution would be to make on-street charging free for all. A combination of free charging for EVs, together with bans on internal combustion cars in city centers would bring clean air and quieter streets. Regardless of the longer-term solution, the problem of price disparity is a real and present danger to the EV revolution.

Even at the higher average rates for electricity, in almost every state and country charging an EV is still cheaper per mile than paying for gasoline. The lifetime savings costs, with the lack of EV maintenance, can also be offset against the running costs. A final added advantage with electric cars versus gas cars is that the price of electricity tends to change very slowly in comparison to gasoline prices which can jump or decrease overnight. Most energy prices fluctuate against the price of oil, coal and other fossil fuels but at a more gradual rate than at the gas pump. As I will explain in chapter 10, the progressive shift towards renewable sources of energy and the reduction in dependence on fossil fuels for electricity production will further stabilize energy prices.

As with all new technologies, competition will drive prices down. With the proliferation of competing Charge Point Operators and smarter navigation showing the predicted cost for your charging in advance, prices are likely to remain stable if not slightly decrease. Charging an electric car and getting 300 plus miles of range for just $10, in the best case, seems like very good value for money, however, it will be a while yet before there is a predictable, consistent rate across the board.

SAVING ENERGY

When it comes to optimizing energy storage, there is a distinct difference between internal combustion cars and electric vehicles. If for example, you filled your ICE car with fuel on Monday morning and then didn't use the car for a week, only to find you lost between 5% and 10% of your gas through evaporation you wouldn't be very happy. The thoughts of your hard-earned dollars just evaporating whenever you weren't using your car would make you change some of your habits about how and when you fuel up. This is what happens with electric cars. Batteries are effectively chemical reactions, charging and discharging moves ions from one material to another for storage and discharge. Many electric cars are never really turned off. There is no start and stop ignition, battery management systems are always monitoring the outside climate to optimize energy storage. For example, a Tesla will warm a battery if it gets too cold to avoid energy loss. Cold weather degrades the charge in a battery that's left sitting around. I'll explain the science behind this in the next chapter but for now, what you need to know is that what you put in will not be what you get out of your car is left unused. Firstly, it's important to understand we're talking small numbers here, remember that over 80% of the energy that's put into an electric car comes out in the form of movement, versus at least 70% of energy wasted from internal combustion cars. This doesn't mean that if we can still save a few pennies with smart technology that we shouldn't. Remember when there are millions of electric cars on the road all these pennies will add up to a sizable energy-saving and have a climate-positive impact.

There are a couple of ways to minimize energy loss in electric cars. The first, and obvious way is to only charge just before you need it. The second is to only charge at off-peak, often cheaper rates. The third is to use smart battery management system software to precondition, warm and heat the battery to optimize energy savings. Some EVs are better than others when it comes to energy management. Green Car reports that up to 2% - 3% of overnight energy loss was experienced with the first model Jaguar iPaces, the first Tesla Model 3s weren't much better. The 2020 covid pandemic saw many cars, including electric, lie idle for long periods. Cars prefer to be driven than spend extended periods idle. Not many people are aware but electric cars also have a standard 12-volt battery, just like a gas car. This is for emergency use and as a backup to keep the software alive in the event the main car battery pack is depleted. Just like leaving your car idle for weeks in winter, this battery can also go dead. Many EVs use this 12-volt battery to power the software for the main battery management system, making it a vital part of the car's health. It is therefore more important than with an ICE car that can simply be jumpstarted, to make sure your EV never completely runs out!

Most cars will go into 'sleep mode' after several days of being idle, however, this doesn't stop the main traction battery pack from continuing to dissipate energy. Maintaining a lithium-ion pack in a full or low state of charge for extended durations is not healthy for the long-term life of the battery. The general rule of thumb is to keep a battery between 20% and 80% charged. Most electric cars can set a charging limit to avoid constantly overcharging the car, and most manufacturers recommend leaving a car that is going to be stored for several weeks plugged into a wall charger.

How much energy will be lost by an idle electric car depends on lots of factors, particularly outside temperature? I once left my car outdoors for two weeks in the middle of winter at Frankfurt Hahn airport in Germany. When I parked up my Tesla two weeks earlier there was still just over 100 kilometers remaining on the battery. However, temperatures dropped to double-digit below, hitting -12 Celsius / 10 Fahrenheit. When I returned just 20 kilometers were remaining on the battery which put me under a bit of pressure to find a charge before I headed home!

Lithium-ion batteries also don't like extreme heat, and the battery management system will often kick in cooling on hot days even if the car is idle. Most electric cars come with apps where you can control the charge level, monitor the battery even when it's parked, and put the car into extended sleep mode. Using smart charging, to ensure a car is never under or overcharged regularly, combined with some active monitoring of idle electric cars can keep your energy loss and any ill effects on your battery to a minimum. At the end of the day, all batteries will not hold a charge forever if unused and in the next chapter I'll cover some of the new technologies emerging to improve longevity. However, what happens if your car completely runs out? If the 12 Volt battery is still operational you can still open the charge port and even if it is not operational there are hacks available to open the charge port, including secret emergency charge port opening buttons often in the wheel wells of cars. If you are nowhere near a charger when it dies completely what can you do? Well in addition to having a tow to a charger there are some mobile charging solutions now operational. I saw these in Munich airport in 2021; they are large batteries that can be towed to a dead car to kick it back to life. Several startups are working on these solutions for breakdown services around the world. And just like the well-known breakdown services for gas cars, there's no doubt this will become standard across the globe as electric cars proliferate.

WIRELESS CHARGING

Nikola Tesla experimented with over the air electricity transfer back in the 1890s with his Tesla Tower at Wardenclyffe, in Shoreham, Long Island. Near field Wireless Power Transfer (WPT) uses induction. Wireless charging has been around for decades and today you probably use it for your smartphone. Research into induction charging for electric cars has been underway for several years and working prototypes are now in use. The Society of Automotive Engineers (SAE) developed a universal standard in 2020 for wireless electric car charging. Known as SAE J2954 this applies to inductive charging systems up to 11 kilowatts.

Induction charging works by creating a magnetic resonance field between a ground-based charging pad, or transmitter, and a receiving pad under the car. Now wireless induction charging is limited to speeds equivalent to a level 2 plug-in charger and below, DC fast charging with wireless isn't currently commercially possible although it's close. Wire coils in the pad on the transmitter and receiving side create a magnetic field, a vibrational resonance that can be transformed into electrical energy. Until recently, induction charging, for any device, not just electric cars, had been relatively inefficient and slow, with significant energy loss in the transfer.

The big advantage wireless charging can bring, would be to eliminate cumbersome and unsightly cables. Wireless charging could also be incorporated into city streets, with charging pads in city parking spaces and car parks. This would avoid the necessity to have to run cables from the car to a charging post. The problem right now, as with a lot of new technologies, is the price point is simply too high compared to just a wall box or post charger. As the technology develops, wireless charging could feature not just in your home driveway, but all over city streets. No user intervention in the charging process would make recharging your car even simpler than going to a gas station. A recent JD Power mobility disrupter study found that over two-thirds of consumers in Germany would be more encouraged to buy an electric car if they could charge it wirelessly. Work is also underway on bidirectional power transfer for wireless vehicle to grid. How far advanced and how close is wireless technology to becoming mainstream?

Several startups are focusing on wireless charging solutions. WiTricity based in the US is working in collaboration with the Massachusetts Institute of Technology (MIT) and is pioneering research into developing wireless charging solutions. BMW already offer a wireless charging option on some of its hybrid models. Currently choosing this charging option from BMW comes with a price tag of over $4,000 making it at least four times more expensive than installing your wall charger. These prices will come down as the tech matures.

Exciting research is also underway into charging cars wirelessly while they are in motion. Israeli startup, Electreon, is experimenting using charging coils embedded in the road. Test tracks have been built in France, Israel, and Sweden to electrify short stretches of road. As early as 2012 carmaker Renault built test tracks to experiment with induction charging while a vehicle was moving. Norway is already working on implementing electric taxis that charge via wireless charging pads beneath street taxi hire stops. This will add just a small amount of range, however, given taxis stop a considerable amount of time during the day, waiting for a fare, it could be enough to ensure taxi drivers never have to go to charging stations.

For now, the distance between transmitter and receiver greatly affects the efficiency of power transfer. For example, 60% of the power can be lost at 1 meter or more. The closer the transmitting pad is the better, the latest electric induction systems for cars can achieve over 90% efficiency power transfer. For wireless charging to be effective now both transmitter and receiver pads need to be very close to each other, known as nearfield charging, this may require some car software tech, like parking sensors, to get the optimized charging position when parking. Aside from the cost of the hardware, on-street and on highway wireless charging solutions will require significantly more capital costs, given streets will have to be dug up and pads installed below the surface. Wireless charging cannot be viewed in isolation, it's aimed to be a complimentary top-up. Another advantage of wireless charging will be when full self-driving and autonomous vehicles become mainstream. Wireless charging would replace some mechanical early ideas whereby charging cables moved robotically into position and connected to a car. Although the technology is still relatively immature, in the future large stretches of highways, with solar panels in the center lane medium, powering battery packs that feed into wireless induction coils embedded in the asphalt could charge your EV as you drive, delivering a highly sustainable and exciting clean mobility solution.

HOW IT SHOULD BE

I started this chapter by characterising the current landscape of EV charging as the Wild West. This may seem a little harsh, but it is the reality on the ground today. Let's look at the path forward, let's look at what needs to happen for our utopian charging world to come to pass. Before I finish this chapter of the book, and as if by divine intervention, I had another strikingly obvious example as to why charging infrastructure connectivity needs massive improvement, this time years apart and in Dublin, Ireland while I was writing this chapter.

When it doesn't work, it sometimes really doesn't work! I was in Ireland in March 2021 and had taken one of our Teslas on hire for a week. I was staying in a ground floor apartment with no access to a charge point. As I wasn't planning to do much mileage, I didn't plan to charge. Near the apartment, there are no fast chargers, just level 2 on street. By the way, at the time of writing this book in 2022, there are still no Tesla superchargers in Dublin. I had a few unexpected longer journeys to do that week and got back to the apartment late one night with just 27 km left on the battery. I didn't panic because I had a few backup plans. Across the road from the apartment is a supermarket that I noticed had just installed a brand-new public charger. On top of that, just about a kilometre from the apartment, is an on-street public charger available for use.

So, I decided that the next morning that I would drive across to the supermarket and charge. When I got to the supermarket this brand-new charger wouldn't work because I had to pre-register and use an app to charge and, once again, the universal charging card I had wasn't yet affiliated with this charging network. This is not the first time this happened to me, the same thing occurred in Amsterdam when I arrived at the hotel that I had specifically chosen because I could see a public charger was available right across the street. Only when we got there did I realise that you have to pre-register with a local city council to use the charger, which took about two to three days before they sent you a charge card charge, in other words, these chargers are useless unless you live in Amsterdam. Imagine having to go to a petrol station and have a whole multitude of registrations before you could fill up your car. I gave up on the Dublin supermarket because I didn't want to register and download another app, instead, I decided to drive 1 km down to the on the street public charger I knew about. When I got the charger a Mercedes S class diesel was parked in the spot reserved for electric cars, meaning I couldn't access the charger, another 'gashole'! I ended up using the 'cable in the window' option at my parents' home. I'm living deep in the new electric world; a newbie would be put off by these experiences. So how can this be fixed…

Connectivity, a simple, consistent, and universal user experience that only routes drivers to working, connected and available chargers. OEMs, charging app providers, charge point operators, governments and service stations need to come together if we want a massive shift to EVs…make it a pleasure, make it simple. Some work on universal roaming charging is underway and a single charging standard, Open Charge Point Protocol, by the Open Charging Alliance (OCPP 1.6 – 2.0). While these standards are more on the physical hardware side, there is still a lot of work to bring a common charging UX to fruition. Here's my top 10 must-dos for charging.

1. Bring the car, the driver and the charger together – let the car do all the work! Not you!*
2. No charging cards
3. No touch screens, choices, buttons or options on chargers
4. One routing standard for all OEMs**
5. Ability to pay with a credit card / contactless for now (although I know longer term this will be phased out once no.1 is fully in place)
6. No more third-party charging apps – a single in-car interface
7. Provide a choice between the fastest route or cheapest route
8. Put simple EV charge point signage on all highways, airports, hotels, etc.

9. Ban hybrids from using public charging points (another story for a later chapter)
10. Share all charging networks – no exclusivity to any OEM

* The solution needs to be that the car does all the 'talking' with the charger and makes the payments (by assigning a payment method to your 'car account' when you buy it - just like Tesla) and then you have zero figuring out to do.

**All OEMs agree and choose a single routing standard (if they are worried about competitive advantage, they needn't be, as, with all new tech, it will eventually standardise and will be solved one way or another anyway, so whoever doesn't sign up will be out of the club).

Solving charging is the second most important requirement for a successful EV revolution. Advancing battery technology is the first. By increasing the density, efficiency, range and cost of batteries, part of this charging problem will go away. The battery is what has made a modern EV possible. The advancements in battery systems for electric cars, in less than two decades, has been incredible, however, in some ways, the story of the battery has only just begun.

CHAPTER 8

BATTERIES

THE LONG AND WINDING ROAD

I was only around 10 years old when the battery leaked all over my friend's mothers carpet. Most pre-teen memories are long gone, however, I vividly remember staying over in my neighbour's house across the street. We were trying to stay up all night and playing around with flashlights that kept going dead. Putting the batteries on top of the radiator for a few minutes 'revived' them, getting a bit more juice for our lights. We left the batteries on top of the radiators and fell asleep. I remember his mother dishing out to the two of us over the brown liquid leaking out of the battery and dripping down the radiator onto the carpet the next morning.

I know now that this was caused by expanding hydrogen in the battery which ruptured the seals, causing a potassium hydroxide leak, or as we commonly call it - battery acid. Warming these batteries sped up the chemical reaction allowing the remaining electrons to produce a current before eventually totally discharging. This unofficial childhood experiment can sum up how batteries work best. When they are cold, chemical reactions happen slower which equals less current, when warm equals more current. Batteries are chemical reactions. You may remember from your science class building batteries with vinegar. Passing ions from one material to the next creates an electrical charge.

Creating an electrical charge like this is not a modern phenomenon and predates the use of electricity. Yes, batteries came before electricity generation and transmission. The earliest batteries or “Wet Cells” were made up of open, separated cylinders containing liquid electrolyte and metal electrodes. These could be “topped up” with water and acid. For those old enough you may remember car batteries being topped up with water. I can remember my grandfather topping up the battery on an old Ford he had. The electrolyte liquid can conduct electricity when dissolved with water. The oldest known battery is thought to be over 2,000 years old and was discovered near Baghdad, Iraq. Scientists hypothesised that the battery may have been used in a form of electrotherapy.

Benjamin Franklin coined the term “battery” in his experiments with capacitors in 1749, however, it wasn’t until an Italian, Alessandro Volta that the first true battery was invented. His ‘Voltaic pile’ was made up of pairs of copper and zinc discs placed on top of each other, each separated by a piece of cloth. This was the first battery to deliver a consistent continuous charge and was able to hold a charge when not in use. Like the combustion engine, the fundamental design of batteries changed little for almost two centuries.

So how exactly do batteries work and why did it take so long before modern electric cars were possible? It’s time for a little chemistry lesson and I’m going to briefly explain just how batteries work. Understanding how they work on a basic level will help you to understand how modern lithium-ion car batteries work and why there is such a race to make them even better. The batteries you use every day in your appliances are known as dry cell batteries. These batteries convert stored chemical energy into an electrical charge. The three core elements of almost every battery are an anode, containing the negative charge, a cathode, containing the positive charge and electrolyte, the substance used to help the ions travel.

The most common elements used are zinc for the anode, a graphite core for the cathode and ammonium chloride paste for the electrolyte. Batteries are chemical imbalances; in the anode, a buildup of electrical ions wants to get to the cathode to neutralize the battery. The electrolyte barrier blocks this transfer of electrons from the anode to the cathode until they are connected at each end allowing electrons to travel to the cathode, your electrical appliance then taps into this flow. Within this circuit 'open' atoms in the anode eventually lose all their electrons to the cathode, turning the atoms into positively charged ions. When the anode is empty of electrons and the cathode is full your battery is no longer capable of giving a charge. That's the end of the road for your battery unless it is rechargeable. Rechargeable batteries operate slightly different. With a rechargeable battery by applying energy, you can move the electrons and ions back to the anode to be used again, however, this cannot be done indefinitely and not all batteries can be recharged, I'll explain why.

The common batteries like AA, AAA, C and D cells are all alkaline batteries. In these non-rechargeable batteries, the cathode is made of manganese dioxide, the anode is a zinc powder, the electrolyte is potassium hydroxide (this is the alkaline part and that liquid that spoiled the carpet!). Alkaline batteries have a high pH, where P is for power and H for hydrogen. In non-rechargeable batteries the electrolyte corrodes the anode, gradually diminishing power, which is why if you heat that dying battery or smack your flashlight you get a little bit extra power for a few seconds because you are shaking up the remaining ions. Non-rechargeable batteries are known as primary cells and rechargeable known as secondary cells. If you try to recharge a primary cell, the charge isn't held in the alkaline battery causing the release of hydrogen gas and possibly even an explosion.

This is where lithium-ion comes into play; most secondary rechargeable cells are lithium-ion today. Lithium is a highly reactive element, and modern batteries use a combination of lithium cobalt oxide as the positive cathode and just simple carbon as an anode. These elements allow electrons and ions to be pushed back and forth between the carbon and the lithium repeatedly, but not forever. Non-rechargeable manganese batteries are relatively cheap and good for low draw devices while lithium-ion is good for high draw devices such as car motors. Somewhere in between alkaline and lithium batteries is nickel-cadmium (Ni-Cd) batteries which can be recharged and are good for smaller devices.

Lithium-ion batteries eventually die, mainly due to internal corrosion. Small crystalline formations grow inside lithium-ion batteries as they age. These are known as dendrites. Dendrites grow inside your battery and gradually reduce efficiency by consuming the liquid electrolyte. Dendrites are like stalactites in a cave, are metallic, and eventually, form little wires that short circuit the battery cell. In modern electric car battery packs, there are thousands of individual lithium-ion cells. The failure of any cell does not destroy the entire battery, it's a redundant design. Indeed, even if half the cells failed the battery pack would still function although you would lose half your charge output and range. The invention of the lithium-ion battery pack fundamentally created the modern electric car today. The first lithium-ion battery was developed in 1985, based on research, in part, by Nobel prize winner John Goodenough in the 1970s and first used by Sony in some products. While the use and advancement of lithium-ion cells have transformed modern electronics from smartphones and laptops to cars, we are approaching the fundamental limits of the technology. Solving the problems of dendrites is fundamental to advancing further.

Research and prototyping into lithium metal cells, new forms of electrolyte and solid-state batteries are advancing at a rapid pace. Promises of a future battery capable of a 1,000-mile range and five-minute charging have been claimed. For now, lithium-ion battery density and output per kilowatt has been the focus for the past decade. Reaching a battery pack cost to make mass affordable electric cars is the EV Holy Grail. Massive achievements in battery technology have occurred in just the past 20 years, however, this is not just down to the elements used but also to modern battery management systems. I'm going to explain how battery management systems work and why they are fundamental to progress EVs and why falling battery production costs will rapidly accelerate the adoption of electric cars; I refer to this as the kilowatt race.

THE EVERLASTING BATTERY

In September 1977, NASA launched the Voyager One probe to explore the outer planets of the solar system. Voyager One has now left the solar system and continues an interstellar journey. Its battery still works, 45 years later. It is powered by a Radioisotope Thermoelectric Generator. This mini nuclear reactor generates heat which in turn is turned into electricity. Unfortunately, this is not the type of battery we want to have in everyday use, as it's radioactive.

Since the development of the lithium-ion battery, most research has been on managing the cells rather than producing a completely new battery technology. This is about to change with massive research, spurred on by the advancement of electric cars, into new battery technologies. Battery management systems are crucial for power management and your safety. Lithium-ion cells can be extremely dangerous. You've heard lots of stories about Teslas catching fire after crashes and I covered in chapter 6 all the facts and fiction about electric car safety.

With so much attention on lithium-ion battery fires, I decided to try a little experiment myself. I don't think we ever really grow up, at least I don't think I have. I tried a childhood-like experiment a year before writing this book with a lithium-ion battery. Don't try this at home! I bought a power tool combo set, that came with several rechargeable lithium-ion 18-volt battery packs. I know the next bit sounds dumb, but I did it anyway. I decided to take one of them apart. Inside there was 12 individual, AA-like battery cells all interconnected together with an electronic circuit board overlaying the entire configuration: that's the battery management system. This battery was fully charged. I decided to conduct two experiments in the garage of my home...

I cut off the battery management electronics. I stripped down some thin copper wire, removed the plastic, and placed it over the positive cathode and negative anode. The copper wire very quickly glowed, like a light bulb, then broke apart. I decided to try something more substantial, I placed two metal wrenches over the positive and negative ends of the battery pack to see what would happen. If you're thinking at this stage all of this sounds a bit dumb, you're right I admit in hindsight this was quite a stupid thing to do. Nothing happened for a few minutes until I began to smell an acrid odour. There was no smoke, however, placing my hand close to the pack I could feel the heat already. A few seconds later, white smoke and sparks started shooting from one of the cells like a firework about to lift off. I panicked as the garage suddenly became engulfed in smoke, pressed the garage door opener and without thinking kicked the battery pack out into my driveway.

An old couple were walking down the street and stopped in bemusement at what was going on. I had to give a friendly convincing wave like I knew what I was doing. The entire pack was now shooting firework-like flames in all directions with an incredible amount of smoke. I was in a total state of panic, as I did not expect this extreme reaction at all. I had to get a water hose to spray the flaming battery pack, which took at least 20 minutes of spraying before it stopped burning and it was still steaming for around an hour afterwards. These were just 11 lithium-ion cells, there are more than 7,000 similar cells in a modern car battery. Lithium-ion cells if not managed properly are extremely dangerous. This is THERMAL RUNAWAY. I got the message.

Eberhard and Tarpenning in the early Tesla days developed the first electric car battery management unit (BMU) capable of taming thousands of lithium-ion cells. BMUs are perhaps the important part of an EV. While most modern devices powered by lithium-ion batteries have BMUs installed, how an electric car's BMU operates is completely different to say a mobile phone. The BMU is an electric car that isn't just focused on safety and performance, but also longevity, whereas mobile phones are constantly upgraded and discarded after two or three years. The actual full capacity of a new EV is more than is made available or visible to the driver. The battery management unit keeps this excess power and range in reserve. This is what allowed Tesla to release more range to Tesla drivers during hurricane Irma in Florida.

As battery packs age the efficiency and range of the vehicle begins to decrease the BMU frees up some of this spare reserve capacity. This means that even though the physical capacity of the battery pack has diminished, the perceived capacity for the driver has little changed. You may think this is false advertising and misleading, however, this secret reserve capacity also protects the pack when power is needed for heating and cooling. Some cells in the pack can be almost fully charged while others can be nearly depleted. This is one of the functions of the battery management system, to make sure that there is uniformity in the charging so that no one cell is over or undercharged. Without this cell balancing, the longevity of the battery would diminish at a much faster rate. Indeed, some cells even die and the BMU effectively removes them from the rest of the collective. The BMU, therefore, separates cells and groups of cells into a series of protective gates; a gate can be shut to protect the wider battery pack.

If the priority of the BMU is to protect cells, the second is temperature control. Charging and discharging lithium-ion cells creates heat. As with my ill-fated experiment above, controlling thermal runaways is vital for safety and performance. Using coolers, heaters and fans the BMU regulates temperature. Coolant strips snake in between the battery cells to provide terminal protection in the event of under or overheating. Of course, cooling the battery requires energy and this is perhaps one of the least efficient aspects of an electric vehicle.

Unlike an internal combustion car, the heat loss is relatively small, hence contributing to an electric vehicle's 80% efficiency rate. Monitoring individual cell health, state of charge, state of discharge and temperature is all part of the BMU's role. The rate and speed of recharging, particularly supercharging, is controlled by the BMU to optimize charge time safely. Therefore, supercharging slows down as you approach 80% or more charged. As with lithium-ion cells, battery management systems are largely reaching the limits of their capabilities. With over the air updates Tesla was able to slightly increase the efficiency of supercharging by releasing software that pre-heated or cooled a battery to the optimal charging temperature as the car approached a supercharger.

These enhancements are reaching the peak of the physical hardware. I'll explain later in this chapter how Tesla and Volkswagen are taking existing lithium-ion configurations and BMUs to the next level. The cost and efficiency of batteries are changing rapidly, making EVs increasingly more affordable.

THE KILOWATT RACE

Gordon Moore, of Intel fame, famously developed "Moore's Law" in 1965. Moore's Law states that the number of transistors per silicon chip will double roughly every two years while the price would halve over the same period. We've become quite used to new technologies entering our lives at a high price at first, before dropping to a fraction in the years that follow, like flat-screen TVs and mobile phones. IBM announced a huge advancement in processing power, efficiency, and transistor chip size of just 2 nanometers, 6 times smaller and 75% more efficient than other chips on the market in mid-2021. Moore's Law has proved true repeatedly.

Unfortunately, the same cannot be applied to batteries. While there has been tremendous progress made in the price of battery cell capacity over the last 20 years, unfortunately, Moore's Law simply doesn't and can never apply to battery technology. The race has been on to develop a battery that costs less than $100 per kWh. In 2010 the price was over $1,200 per kWh. The price for the latest Tesla Model 3 battery is around $130 per kWh. Although there has been a 10-fold cost reduction in 10 years, it has not followed Moore's Law, and the pace has begun to slow as we reach the limits of lithium-ion. The $100 threshold is significant as it will see electric cars become cheaper than their gasoline equivalents to build.

The cost of the battery in an electric car is the single most expensive item. Both Tesla and Volkswagen's new battery configurations, as I'll explain below, are already promising a below $100 per kWh battery by

2023. Reducing battery costs is not just about smarter technology and better manufacturing processes, but also smart sourcing materials and minerals.

Compared to ICE manufacturing, electric is in its infancy. Economies of scale will come as the world shifts to electric, which will drive battery costs down even further. Therefore, there will be economic reductions in the cost per battery pack, along with more technological advancements in management systems, improved mining techniques and metal supply chains. I will explain later in this chapter where all the elements come from and just how they get into your car. The reason why almost all the EVs produced on the market so far have been large luxury category cars is because the price of the battery is still so expensive. The manufacturers wanted to try and make a margin on the more expensive cars. Cheaper batteries have meant cars like the Tesla's Model 3 and Volkswagen's ID 3 are now possible, however, we are still not at mass affordability.

Before I get into the new research and technologies, let's look at just what else can be done with existing technology to lower prices and squeeze out more range. Starting with Tesla, their 'Battery Day', on September 22nd, 2020, did contain some of the usual 'Musk hype' but the various announcements signal more major milestones in the accelerating electric revolution. Improving battery production methods and the future metal composition of batteries will deliver more efficient and higher density batteries for faster, longer-range cars.

From mining to production, recycling and metal casting techniques presented by Musk promise to deliver 33% lower lithium production costs for example. Incorporating the battery pack to become the structure of the car itself (instead of a separate battery within the floor) will allow for greater density and the possibility of even further range. This alone promises 35% floor space capacity saving. Overall, the various announcements promise a 56% reduction in the key electric car measure of $ per Kilowatt hour, promising the future $25,000 EV. From 2018 to 2019, alone, the drop in battery pack production cost has been 13%. With announcements like these, we will continue to see further drops as advances in research and development combined with economies of scale take hold. The announcement of Tesla's new 'tabless' 4680 cell promises higher energy density and improved power to weight ratios while streamlining manufacturing. These thicker cells, 46 millimetres wide and 80 high, look like the mini soda cans you get on an airplane.

I'll cover some of the startups that are promising a battery revolution in the years ahead but next let's look at Volkswagen's attempt to win the kilowatt race. Volkswagen is in the middle of metamorphosis to become a 100% electric car maker. It is incredible to think that not long ago, in 2017, Volkswagen was ordered, by a US Federal Judge, to pay $2.8 billion in criminal fines for rigging diesel-powered cars to cheat on government emissions tests by hiding the true level of the poisonous gas nitrogen dioxide (NO_2), emitting from VW tailpipes. The software used to cheat the system hid levels of NO_2 that were 40 times worse in the real world than reported. 'Dieselgate', cost VW over $30 billion in fines, penalties, and settlements. VW sold its diesel cars with gusto in the 2000s — in some cases advertising 'Clean Diesel' Technology. There was never anything 'clean' about diesel and the sooner it is confined to the history books, the better for our air quality and climate crisis. Fast forward to today, VW held its 'Power Day' event in February 2021, unveiling a strategic roadmap to go 'all in' on battery development. Remarkably like Tesla's 'Battery Day', - battery technology is VW's focus. Volkswagen released its much-anticipated ID.3 electric car. The Golf lookalike is the first VW electric car built from the ground up — comprising the floor plate battery pack and electric motors configuration known as the 'skateboard'. The ID.3 follows the previous half-hearted attempt, the eGOLF. This was an overweight, overpriced, under-ranged Frankenstein monster that has now been very much replaced by the ID.3. VW have a great reputation for making good, reliable solid cars. My first car was a 1988 VW Golf, I loved it, drove the hell out of it and it never let me down.

Following the shock rise of Tesla, VW is now on a real path to becoming a fully electric car maker and has ambitions to be number one. Asleep at the wheel for two decades while Tesla rose, VW is making a massive, concerted effort to turn electric. Producing an electric car is entirely different to building an ICE car. Despite all their years of proficiency building ICE cars they had, until now, limited capacity and expertise to replicate Tesla's one million EV head start. To do this, VW needs to redesign old plants, build new plants, remove legacy inhibitors, and most importantly build batteries on a massive scale.

The VW announcements confirm they are serious about becoming a global electric car goliath. In 2020 VW announced it would invest over $80 billion in electric car technologies, double the Dieselgate cost. Mastering battery production is the number one priority to achieve its goals. The big stories from its Power Day were its plan to reduce battery costs by up to 50% and be capable of a battery recycling rate of up to 95%. VW didn't just announce an 'economies of scale' plan to reduce battery costs, but new production methods and technologies. They include the development of six 'Gigafactories' with a total production capacity of 240 gigawatts of battery output to be operational by 2030. To reduce the cost and complexity of battery cell and pack production, VW plans to integrate battery management systems (mostly the essential thermal controls) into the cells themselves instead of separate external battery management systems—known as a unified cell architecture. By breaking, and going below, the $100 per kilowatt-hour threshold, VW aims to make the 'affordable-by-all' electric car. VW's Group CEO, Herbert Diess, said "The transformation will be bigger than anything the industry has seen in the last century...it will be fast, it will be unprecedented". Squeezing the last bit of efficiency out of existing lithium-ion technology systems is just the beginning, let's now look at the research and startup frenzy taking place in battery technology and what the future may bring.

SOLID-STATE STARTUP FRENZY

As the pace of the electric revolution quickens, there has been a virtual gold rush into this, soon to be vast, electric industry - it's almost all about the battery. Improving battery density, the amount of energy that can be contained in the same volume is the focus of startups and research. So, what's the next big thing after the lithium-ion cell?

Many of these startups focus on a new technology known as 'solid state'. You're going to hear a lot about solid-state batteries in the years to come but what are they? As the name implies instead of using liquids or gels for electrolytes (which can form dendrites over time) solid-state batteries use solid electrodes and solid electrolytes. This means the electrons moving from the anode to the cathode don't pass through a liquid medium. Inside a conventional lithium-ion cathode the particles are made up of lithium-containing metal oxide, such as Nickel, Manganese Cobalt oxide, known as NMC. These elements effectively hold the structure of the cathode together when the battery has been fully discharged. When the battery charges, the lithium particles leave the cathode, travel through the liquid or gel electrolyte and enter the anode ready for discharging again. Carbon atoms in the anode temporarily hold on to lithium particles until it's time to be discharged and sent back into the cathode.

Eventually, this constant moving back and forth degrades the battery, as I've explained, however, in solid-state batteries, this is very different. In solid-state, there are just two layers instead of three. There is only a positive cathode in a solid-state battery with a ceramic separator to the electric contact. As a solid-state battery charges, the lithium electrons leave the cathode and accumulate in between the ceramic separator and the electrical contact plate. An anode is therefore formed in between this gap and is pure metallic lithium. This allows the energy to be contained in a much smaller volume, therefore increasing the energy density. Solid-state lithium batteries offer the potential for ultra-fast charging and increased safety by reducing fire risk. With more efficient thermal control, less work is required for battery management systems, cooling and heating components making the battery pack even simpler and more cost-effective.

Solid-state batteries are to electric cars what nuclear fusion will be to energy production, transformational. Like nuclear fusion, the solid-state theory is not new, indeed Michael Faraday made early discoveries in the mid-1800s on which today's batteries are based. The solid electrolyte in these batteries can even be made from glass. Solid-state doesn't just deliver higher density, because it's a more efficient use of volume it means batteries can be smaller and therefore a lot lighter, further improving the efficiency and range of EVs. Some estimates predict that solid-state can at least double density and range, and because the formation of dendrites is controlled, the longevity of the battery could also increase by a factor of two or more. Increasing the number of lifetime charging cycles is equally important to ensure longevity.

Achieving a range greater than an internal combustion car, on a battery that effectively lasts for decades and at a price that's affordable to the masses would transform the electric car revolution into an "overnight" phenomenon. Internal combustion cars by comparison would look like ancient technology and their likely phase-out would be rapid. Italian supercar maker, Automobili Estrema, unveiled a solid-state hyper car prototype in mid-2021.

Different automakers have made differing claims about just how close their solid-state technology is. For example, luxury automaker Bentley predicts its range of electric cars will be solid state-driven soon, while Volkswagen believes this technology is at least a decade away. In Japan electric laggard Toyota has formed a consortium, named LIBTEC, with Honda and Nissan, to advance solid-state production. Tesla's battery supplier, Panasonic, is working on solid-state technology and predicts the technology will be available in electric cars by the end of the 2020s decade.

As ever, it's mostly not about the incumbent legacy manufacturers, startups are driving a virtual feeding frenzy on solid-state technology. One startup gained international media attention in 2020 with its announcement of a breakthrough in solid-state battery technology capable of charging a standard EV to 80% in just 15 minutes or less. Quantum Scape's announcement drew some sceptics, but the company has defended its claims, slating 2024 for full production. Volkswagen, along with Bill Gates, has invested heavily in the San Jose-based start-up.

Sila Nanotechnologies has raised $590 million in 2021 and has plans to build a factory with triple the capacity of Tesla's Nevada Gigafactory and develop a commercial-grade silicon anode, something researchers have been working on for years. Going one step further Israeli battery company StoreDot claimed breakthroughs for extreme fast-charging battery (XFC) technology in early 2021. StoreDot's technology is based on silicon nanomaterials replacing graphite to control battery swelling as ions fill the anode. The company claims the technology will deliver a full EV charge in just five minutes. I would argue this level of focus on extreme fast charging is not necessary, given 95% of use cases don't need ultra-fast charging for everyday mobility. A host of other startups are working on similar technology to replace graphite in the battery through materials science research, with silicon looking like the potential winner material. Relative to lithium and cobalt, silicon is abundant and one of the easiest to mine minerals on Earth.

Mature companies like Samsung have developed a solid-state prototype using silver carbon to create the anode and claims the battery will be capable of delivering an EV range of 800 kilometers (500 miles) on a single charge. While Toyota has been very slow to come to the electric vehicle party, they do have over 1,000 patents registered on solid-state battery technology. Toyota has claimed that its batteries will still operate at 90% capacity even after 30 years and be fully chargeable in just 10 minutes. Despite their hybrid focus, Toyota has been working on solid-state batteries for almost 10 years.

In parallel, big auto players are also investing in partnering up with some startups. As mentioned, Volkswagen invested in and owns 5% of Quantum Scape. BMW formed a partnership with Solid Power to develop solid-state batteries for cars and while Dyson's electric car never made it into production, they have also acquired solid-state battery startups like Sakti3. Hyundai is working with Ioniq Materials in a similar pursuit.

It's not just startups and the car industry, tech behemoth IBM is also researching solid-state. The science of quantum theory analyses particles at a subatomic level to understand their interaction with each other. IBM, together with German carmaker Daimler, is researching, at the atomic level, the chemical interactions between elements in a battery to better understand and improve capacity, charging and discharging. National governments are also getting in on the solid-state race. Japan and China have gone head-to-head in funding research and development. Although solid-state batteries are expensive to manufacture today, it is expected that with economies of scale, the price will drop rapidly.

Solid-state batteries are still not in use in consumer electronics or any true commercial form. Getting the electrolyte material correct, to maximize ion transfer efficiency is part of the continuing research and development. However, with promised higher density, lighter weight, longer range, higher safety and longer-lasting life, solid-state batteries offer game-changing potential not just for cars, but energy storage and future consumer electronics. Your smartphone battery, for example, could last at least 10 years. If all goes to plan, and the technology becomes commercial, the 2030s decade will witness a revolution in mobility, electronics and energy storage, hopefully mostly from renewable energy sources. Before this can happen, production techniques need to significantly develop, for example manufacturing a solid-state battery requires an extremely dry environment to prevent any moisture from entering the structure of the battery. On top of that, supply chains for mass production will require multiple different metals and elements to be available on a massively increased scale. This brings us to the next big part of the battery revolution, supplying the materials. Just like oil, these materials must be taken out of the ground.

MINING

Some say electric mobility is just moving from one bad substance to another. Oil to lithium. Lithium is no saint, however, it's no comparison to oil. That would be like comparing the flask to the hot liquid that's in it. We don't burn lithium; it stores the energy we produce. Batteries are far better than refining and burning oil as a means of energy transfer. The minerals and metals in batteries need to be mined and mining itself produces carbon dioxide, unwanted byproducts, and is often carried out in geopolitically sensitive regions.

The primary mined elements for batteries include nickel, aluminium, phosphorus, iron, copper, graphite, lithium, cobalt and manganese. Global demand for these elements is expected to grow by between 10 to 15 times by 2030. With this huge growth comes multiple challenges, from mining, to supply chain, to quelling geopolitical tensions. These raw materials often come from poorer countries where mining activities are essential income drivers to the local economies.

Over 50% of the world's cobalt supply comes from the Democratic Republic of Congo, for example, a region that has experienced on and off political instability for the past 60 years. Cobalt in lithium-cobalt batteries is currently superior to other forms of lithium-ion batteries because they offer several advantageous properties, for example, a high energy density, relatively low weight, and better thermal stability. Cobalt also increases the longevity of batteries and is highly recyclable (I'll cover that later). Although cobalt is a wonder material, manufacturers are trying to reduce the percentage in their batteries and limit their dependence on international cobalt supply. Two prime reasons for this, firstly cost, cobalt is a very expensive element at around $33,000 per ton (that's twice the price of nickel) and secondly, ethical concerns around mining, particularly in the Congo with allegations of child labour, corruption, and harsh working environments. Organizations, supported by EV manufacturers, like those for fair trade coffee, have been born. For example, The Fair Cobalt alliance, the Responsible Cobalt Initiative, and the Clean Cobalt Initiative.

Even before solid-state batteries, and even after they come into production, achieving a low carbon mobility future is dependent on six metals. In addition to these six metals such as lithium, copper, and nickel are other Rare-Earth Elements (REEs), important not just for electric cars, but also for energy storage and clean energy. Indium and gallium, for example, are used in the coatings of solar panel photovoltaic film. Establishing a reliable supply chain, at the right price point, which is also massively scalable, is essential to complete the electric revolution. So how abundant are these metals? How reliable is the supply? And what needs to be done to meet the massive EV battery demand predicted?

Starting with lithium, soaring EV demand has given rise to the creation of new lithium mines. Lithium comprises about 6% of the battery and is the 25th most abundant element on earth. In Australia for example, where over 50% of global production occurs, 6 new lithium-ion mines have opened since 2017, increasing production capacity by over 20%. Even with increasing EV demand, lithium is a relatively abundant element when compared to some of the other battery metals and increasing production is driving down prices. Lithium is found in rock and clay deposits, and it is also dissolved in briny water. Lithium has been dubbed ‘white oil’ and apart from Australia, just three countries control much of the remaining world supply, Chile, China, and Argentina. The lithium triangle is a region in South America that borders Argentina, Bolivia, and Chile. It is thought that over 50% of the world’s lithium reserves are contained there, which is roughly the size of the state of Arizona or Italy.

In Europe, recent lithium discoveries in Portugal offers the continent the potential to significantly supply the region’s demand. Localizing the supply of lithium cuts down on the overall supply chain and additional carbon footprint. Sourcing new lithium deposits is now a key focus area for the leading EV markets of Europe and the US. In the US, with the equivalent of one Gigafactory for battery production being built every four months in contrast to the equivalent of 1 being built in China every week, there are concerns about the continuity of lithium supply. Combine this with the potential geopolitical risks by the US becoming too dependent on China for electric batteries. While the US is estimated to contain almost 7 million tons of lithium, as of mid-2021 zero per cent of this lithium, or nickel, is produced in the USA for EV batteries. The world’s known lithium reserves are predicted to last for at least 200 years, but the US produces just 1.2% of the world’s lithium. Developing a lithium supply chain domestically in the United States is, therefore, a crucial strategic endeavour to avoid the US becoming overly dependent on foreign supplies.

The global lithium market is currently valued, in 2022, at around $50 billion and is expected to grow to at least $200 billion by the end of this decade. In the UK, there is research on deep boreholes, tapping into geothermal waters beneath the Earth's surface to detect lithium. The lithium could then be extracted, like salt production from a salinization plant. Concept technologies to then reuse the hot water pumped out of deep boreholes to power the actual lithium extraction process would make the entire endeavour highly sustainable. So, when it comes to lithium the problem is not really if there are enough reserves, it's more where they are, and who is further ahead in mining and production.

China is by far the leader when it comes to battery cells output today. Indeed, until recently China effectively mandated automakers to use Chinese cells. In Tesla's Shanghai car plant, Chinese-made Tesla Model 3s contain Chinese domestic-made batteries by Chinese CATL. In the US, Japanese Panasonic supplies car batteries to Tesla as there are effectively no domestic US-owned companies producing electric car batteries as of mid-2021. LG is working with GM on their electric ambitions. Other foreign players like the South Korean owned SK innovation (who in May 2021, settled a long-running legal dispute initiated by an international trade commission centering on intellectual property rights) is building new battery plants in Georgia capable of supplying up to 300,000 EVs per year. When it comes to the other key elements, for example, cobalt, manganese, nickel and carbon (graphite), most of these are mined in China, Indonesia, South Africa, and Congo. Most of these minerals are then shipped to China, where over 80% of the world's raw mineral refining takes place, further demonstrating the overdependence on a single country for battery production. All these factors give China the global leadership in battery cell production with almost 80% of the world's battery output. China, Japan, Canada, South Korea, and Germany lead the United States in the lithium-ion battery supply chain. With the Biden administration's push into the new green economy, there is going to have to be a structural shift. The United States will need to do a similar transformation to the oil fracking industry which reduced dependence on Middle Eastern oil when it comes to battery elements and production. The good news for the US is that there it is estimated to have more lithium reserves than China, the problem is the lack of infrastructure to date. With the tipping point now reached, and major American automakers committing to an electric future, America is very much now in lithium catchup mode, a decade behind China's lead in battery production and mining.

In 2021 the early signs of a real battery supply shortage hit Tesla with Musk indicating the Tesla Semi-truck project will be delayed due to battery supply concerns. Tesla consumed more lithium than BYD, VW, Renault, & Audi combined in 2020 and has secured the rights to mine lithium on 10,000 acres of land in Nevada. Tesla's lead already gives it a battery cost advantage over competitors which it is expected to hang onto until 2030.

Next to lithium, copper is a starring metal in electric cars, not just for the batteries, however, also in electric motors, wiring and circuitry. Concerns about global copper supply against rapidly growing demand have driven the price to all-time highs in recent years, more than doubling since the year 2000. A conventional car contains 18-49 pounds (22 kilograms) of copper, an electric car contains at least five times this amount. Like peak oil (the point at which global oil supplies reached their maximum output and begin to decline) peak copper has been debated to be as early as 2035 or as late as 2100. Unlike oil, copper can be recycled, and it is estimated 80% of all the copper ever mined is still recyclable, estimated to be 550 million tonnes or at least 20 years of today's copper mining production.

Copper is an abundant mineral in the Earth's crust but there are only limited concentrations where extraction is economical. Chile, Australia, Peru, Congo, Zambia, Mexico the US and China account for over 50% of global production with the remaining production widely spread amongst 52 countries. Given this wide dispersion, we are not as dependent on any one country, unlike lithium. So, copper has more of a production capacity issue than world reserves and given its recyclability, it is a highly sustainable material. To take an EV lead, predictable copper and lithium supply is the number one priority.

Just as oil was a major influence on American foreign and geopolitical policies, battery material supply is likely to become center stage for world governments as the electric revolution accelerates. When it comes to the environmental impact of mining, yes, it is far from perfect, however, a recent study by Transport and Environment concluded that the impact of mining and the raw materials needed to make an electric battery, considering recycling, demonstrate that EVs use significantly less raw materials than a fossil fuel car. For the battery mining, 'gold rush' careful navigation of the demands to reduce carbon emissions with the protection of the natural environment is required in the years ahead.

SWAPPING

At the early stages of the electric revolution, from around 2005, battery swap technology gained a lot of momentum. Touted as a solution to slow charging speeds and range anxiety, the technology offered drivers the ability to swap their depleted battery for a fully charged battery at a swapping station, almost as quick as filling a car with gasoline. A car would drive into a swapping pod, the used battery would be mechanically removed from beneath the car while a new battery would be automatically and robotically inserted. In these cases, the entire battery pack was removed and replaced in less than 10 minutes. These technologies emerged before the advent of widespread supercharging, higher battery densities and longer EV range.

Several startups that pioneered the technology fell by the wayside. Tesla demonstrated the technology back in 2013 but later dropped it to instead focus on its supercharging network. It could be argued that the establishment of large-scale battery swapping is more complex and expensive than the installation of superchargers. On top of this, subscription and leasing models need to be in place, and you are never guaranteed what condition the battery is in when your old one is swapped out. Israeli startup, Better Place, was an internationally funded battery switching company. Its first station opened just outside Tel Aviv in 2008 and offered drivers of Renault Fluence electric cars the ability to swap their batteries completely automatically in just a couple of minutes. At its peak, there were over 20 battery swap stations. The company planned further deployments in Denmark and other countries but unfortunately fell into bankruptcy in 2012 having spent almost $1 billion developing the technology. This was not a new concept. An exchangeable battery service was first proposed as early as 1896 to overcome the limited range of early EVs. GeVeCo launched an exchangeable battery service for light electric trucks owned by the General Electric company. The service ran from 1910 to 1924 before ICE cars began to dominate.

Today, China NIO has rolled out electric models with swappable batteries. NIO's model is to allow customers to buy a cheaper electric vehicle and then lease the battery via a subscription service. The model is more to offer customers the opportunity to upgrade their batteries at a future date for newer potentially higher capacity batteries.

This raises an interesting question about the potential longevity of electric cars. Given the few moving parts compared to internal combustion, the battery is the only major component that will wear out. It could be argued, especially with over the air software updates, that electric car chassis and body could essentially last indefinitely with battery upgrades. Unfortunately, this is not the path that EV makers are taking. Battery switch technology has been largely superseded by supercharging networks. On top of this, car manufacturers are increasingly making it more difficult to switch their batteries.

With Tesla's, and now Volkswagen's, new integrated battery modular design, the battery isn't just in the floor plate, it IS the floor plate. A focus on battery recycling rather than swapping is likely to be the future. This hasn't stopped newer swapping startups. One such startup is San Francisco based Ample. Instead of an entire large battery pack being replaced, Ample focuses on smaller modular battery units that can be added and removed. The company launched pilots with Uber, where some test vehicles have been equipped with Ample's technology.

With the latest direction of EV makers to integrate the battery more into the vehicle, making it less swappable combined with longer life batteries, supercharging and future solid-state batteries it is difficult to see a world where battery swapping stations prevail. Like hybrid cars, swapping probably had its place in the last decade but doesn't have a future in the decades ahead. In certain use cases, for example, high use taxi fleets or high use last-mile delivery fleets or even long-haul trucking, battery swap technology may have some niche future, however, battery technology has advanced so much that we've passed the point where this is no longer necessary for passenger car use.

RECYCLING

Batteries are not discarded in landfills. There has been a lot of 'battery bullshit' in the media surrounding electric cars. Batteries that are no longer viable to power an electric car end up in one of two places. Firstly, they are deconstructed, and their raw components recycled, secondly, they get a second life as less demanding energy storage.

I'll cover the second life of batteries and energy storage later, first let's look at recycling. Multiple new innovative startups have come on the scene to tap into the new end-of-life battery market.

Much of the materials can be recovered like cobalt, nickel, lithium and manganese. Common parts like copper and aluminium enter well established, decades-old, recycling processes. Some media reports depict a future where millions of car batteries are piled up suddenly with nowhere to go. Although the battery recycling industry is still relatively small, the market will grow as EVs age. If you assume a 10-year life for an EV, you're looking at around at least one million EV batteries per year requiring recycling by 2030. This would of course exponentially grow as older cars and batteries begin to reach the end of life, however, it is mainly a problem for the next decade.

This said, recycling is already happening today. Nissan is using old Leaf batteries for use on the factory floor in guide robots ferrying parts around. The EU is working on mandates that EV manufacturers recycle their old used batteries, for example. French carmaker Renault is recycling all its electric car batteries. Volkswagen is planning to do the same and has opened a dedicated pilot recycling plant in Salzgitter, Germany. Currently, the dated EU Battery Directive from 2006 stipulates that at least 50% of batteries must be recycled, but that includes the casings, plastics and housing making it a relatively easy target to hit.

Recycling the cell metals, processing them for re-use in new batteries is the bigger challenge.

Many countries already have extended producer responsibility (EP), meaning OEMs could be ultimately responsible for retired EV batteries. So, what happens to an old battery when it heads off for recycling? In Volkswagen's pilot plant the battery is firstly completely discharged, any remaining charge is spent before deconstruction commences. Residual energy in these batteries can also be recovered by discharging them into battery storage units on-site in the recycling center. The battery pack is then easily disassembled and broken down into individual battery modules, each containing hundreds of individual battery cells. The cells are shredded into granules and then dried to remove any of the liquid and gel electrolyte. Special separators then sieve the various metals including lithium, nickel, manganese, copper, cobalt, and the aluminium cell casing. Magnetic separators then collect the different metals by weight. Any plastics, such as housing or cable coatings are separated from the metals by the magnets. The various metals are ground down into powders for refinement, preparation, and reuse.

Former co-founder and Tesla CTO, J. B. Straubel is now CEO for Redwood Materials, a start-up aiming to develop technologies to recycle the future millions of used batteries. Redwood uses a combination of methods to extract used elements from batteries, pyrometallurgy and hydrometallurgy. With the pyro method, unwanted plastics and organic materials are burned from the battery while with the hydro method, lithium-ion cells are soaked in a strong acid to separate the metals. Redwood's technique uses any remaining residual energy from the old batteries to power the heating process. While battery recycling today can recover as low as just 5% of the valuable materials, Redwood claims it could recover between 95% and 98% of the nickel, cobalt, copper, aluminium and graphite material.

Another start-up, Canadian Li-Cycle, uses a different approach to Redwood, by focusing on the hydro method to extract the various materials. Battery Resources claims to recover over 90% of the battery through their 'closed loop' system with a combination of hydro and direct recycling 'rebuilding' the cathode materials at an atomic level. In Europe, Elemental Holding is constructing a €182 million dedicated facility for EV battery recycling in Poland. Part-funded by the European Union, the facility will be one of the first in the world dedicated to EV battery recycling.

Tesla itself announced in 2020 its own plans to develop a battery recycling system which they claim will be quite different from other automakers' approaches. It is thought Tesla will be developing its closed-loop battery recycling process; currently, Tesla relies mainly on 3rd party recyclers.

Regardless of the process before any of these recycled metals can be reused, further refinement and processing are required. New techniques, known as direct recovery, aim to recover cathode materials without destroying the crystalline nanostructure (created during the original lithium preparation). The industrialisation of battery recycling for use in new batteries also offers to reduce the overall carbon footprint of electric cars, cutting down on the 'mine to wheel' CO_2 cost. According to the Institute of Electric and Electronics Engineers, IEEE, there are estimated to be over 100 companies worldwide already working on or planning to recycle lithium-ion batteries. Battery recycling will be a very lucrative, enormous business estimated to be worth as much as $30 billion by 2040. Recycling is a problem the world can and is solving at the earliest stages of this electric revolution, it bodes well for a sustainable future.

SECOND LIFE

EV batteries require a high-power output and fast recharge capability to power heavy cars at speed and for long durations. As a battery degrades it loses charge and maximum output, eventually becoming no longer viable for everyday electric car use. This does not mean the battery is finished forever. Used EV batteries can have a second life as storage in less demanding 'retirement'. Ranging from storage at charging stations, home energy storage fed from renewable sources like solar, backup power packs as emergency power, even powering elevators and refrigerators. Even when these batteries can no longer sustain a car they can still collect and discharge power for another 7 to 10 years.

Let's take a typical 70 kW EV battery; at its peak, when brand new, this single battery would hold enough power to meet the energy demands of an average American home for more than two days. America's electricity use for home use is the highest in the world. In the UK a 70-kilowatt battery could power a home for almost a week, while in China this battery would power an average Chinese household for almost 20 days. So even if this average EV battery has lost half of its capacity, it will still hold enough to power a US home for one day at least. Moreover, the demands from home electricity use are far less extreme than powering a car which draws energy fast for quick acceleration and long durations. This means during a battery's second life; degradation happens at a much slower rate. It is estimated that there could be over 3 million used electric car batteries by 2025. Combining these used batteries with renewable energy would lead to a highly sustainable and productive long lifespan for an EV battery from mine to ultimate recycling.

Given China's already strong lead in electric vehicles, it is predicted China will become a world leader also in second life usage. There has been some anti-second life rhetoric from industry insiders. They argue that recycling a battery and reusing much of its minerals for new batteries is more efficient and ecologically friendly than hanging on to batteries for years and dealing with the problem later. Tesla which already manufactures its Powerwall and Power Pack stationary battery packs prefers to recycle rather than reuse old batteries. One of the reasons Tesla prefers recycling is obviously to sell new power packs but also because, as of 2022, Tesla doesn't have many old battery packs coming back for recycling yet. It will be at least another 5 to 10 years before a significant volume of used Tesla batteries becomes available.

Either way, there is a place for both recycling and re-use. With a less frequent and shallower depth of discharge cycles, using EV batteries for home storage is perhaps the most applicable use case. Not all EV batteries will be reusable, it depends on the ultimate percentage of charge capable capacity after it had a life in an EV. Second life EV batteries could cost as low as just 20% of the original cost of a new battery, at around $40 per kWh. Reusing EV batteries improves battery economics, sustainability and reduces the lifetime carbon footprint. There are challenges before this becomes the reality, namely the challenge of falling new battery costs versus the cost of reconditioning old. Also, there is still not a universal standard battery for electric cars with many manufacturers pursuing their configurations making it more difficult for reuse in some cases for now.

STORAGE WARS

The advancement of lithium-ion batteries for the automotive industry has spilt over into the energy industry. With the welcome growth of electricity generation from renewables, like wind and solar, demand is accelerating to find cost-effective storage solutions. The sun doesn't always shine, the wind doesn't always blow and when it does it can produce electricity at unwanted, low demand times. Storing excess energy from renewables for use at peak demand periods is an essential component for a sustainable energy future. In chapter 10, I'll cover how energy is stored by some new weird and wonderful innovations. Used EV batteries for stationary storage could exceed 200 gigawatt-hours by 2030. Even before EV batteries reach the end of life, they are effectively energy storage on wheels. Vehicle to Grid (V2G) solutions are evolving fast and could offer EV owners the opportunity to effectively charge for free. I'll cover V2G in more detail in chapter 10.

Using used batteries for storage doesn't have to just be because you have solar panels, for example, a customer might install so-called 'Behind the Meter' (BTM) storage to reduce electricity costs. BTM works by loading the battery at off-peak hours. This is the same technology as smart charging that I covered in chapter 7, it's just that the battery is hung on a wall instead of residing in the floor of a car. You can bring down the cost to charge your car at home considerably if you can charge up a wall storage battery at off-peak charge rates, then charge your car at even a faster speed than you charge the storage battery when you want at any time.

The cost of battery storage is relatively expensive compared to other methods, which I'll explain in chapter 10, but it can be less expensive than power outage consequences when used as an emergency supply. Like Tesla's installation of emergency power storage in Australia following outages in 2020/21 or the Texas winter storm, when consumers were left without power, and people even died in their homes due to freezing freak temperatures. However, given the price per kilowatt to store energy in batteries, it is not a cost-effective solution for a long duration. As battery cost per kWh improves, especially if solid-state batteries become reality, commercial large scale energy storage may become more viable and widespread.

Industrial grid and battery energy storage solutions are not just a Tesla driven market. Big players, like GE's energy Reservoir, for example, provide industrial-scale storage solutions for backup power that have been in operation for decades in data centers, server installations and vital businesses where uptime is paramount. What's different with the large-scale energy solutions that we're now seeing is that the advancement of lithium-ion batteries and the cost per kilowatt has reached such a point that they can become a reality on a much wider scale.

Battery energy storage also offers more resilience to energy grids in developing countries. Global demand for battery storage is expected to reach almost 3,000-gigawatt-hours per year by 2040, enough power to run all of New York City for at least nine months on batteries or about half of all renewable energy produced around the world in a day. In the final chapter of the book, I will explain how energy storage is becoming a big and exciting emerging market. Some of the valuations which have been placed on Tesla are, in part, because of its energy storage lead with its battery packs and Powerwalls. With Tesla's solar division, for example with its solar roofs, and energy storage solutions combined with electric cars, they can offer an all-in-one package.

For this reason, some investment pundits predict Tesla will continue to rise in value and continue to have the edge over other OEMs who just produce cars. In many parts of the world, almost 1 billion people don't have access to reliable electricity in 2021. Battery storage solutions combined with renewable energy generation offer power to remote regions not connected to electricity grids.

Not only do these solutions improve access to electricity, but storage also helps reduce the growing carbon output of developing nations. Storage can provide electricity for cooking, for example, almost 40% of the world does not have access to clean fuels. Providing access to affordable reliable and sustainable energy for all is one of the UN's key sustainable development goals (SDGs), battery storage can play a big part in the solution.

The humble battery, which for years existed in a much-unevolved state, has in just two decades, rocketed to create entirely new industries. It's not just the storage and the electric car revolution that has been helped by the lithium-ion wonder, however, all consumer devices and a whole new world of mobility have too, bringing a revolution to everything that moves.

CHAPTER 9
ALL THAT MOVES

BEYOND THE CAR

The electric revolution has spilt over into other forms of mobility and our daily lives. Falling battery costs, increased power density and improving battery management sophistication has spawned new mobility possibilities. Electric kick scooters, mopeds, drones, even flying cars and electric commercial aeroplanes are all now possible, and in some cases, already a reality. Advancements in battery technology, apart from the electric car, may have gone largely unnoticed. Just walk into your local hardware store and you'll see everything seems to be going electric now. I was recently struck to see some public park gardeners using weed hackers and hedge trimmers that were electric. I recently replaced an old petrol push lawnmower with an electric 18 Volt dual battery lithium-ion powered mower. This technology spillover has been happening in part because of the advancements in batteries led by the electric car revolution.

It can be argued that lithium-ion advancements have even changed the way modern society views car ownership and personal transport. So-called Mobility-As-A-Service (MAAS) has created multi-billion-dollar industries in on-street electric scooters, for example, made possible by lithium-ion batteries delivering 20 plus miles of range. On-street electric bicycles have also become popular in many cities, again due to lithium-ion advancements. MAAS is disrupting the very concept of owning a car. The use of shared mobility platforms is growing in cities worldwide. Micro mobility, car sharing, car rental and mid-term leasing will replace direct car ownership in many cases. This trend is accelerating. On top of this, demand for a fully connected experience has been accelerated during the Covid era as consumers demand a clean contactless service and now also preferably electric. In the next and final chapter, I'll explain how the world is making the transition to sustainable green electricity to power this electric revolution, but first, let's look at some of the incredible possibilities now reshaping the world of propelled movement for the first time in nearly a century.

eVTOL

VTOL, Vertical Takeoff and Landing is not new, fossil fuel-powered helicopters and military jets have had this capability for decades, but replace combustion with batteries, and you get eVTOL. Remote control battery-powered drones became popular in the early 2000s and began to find commercial uses on a wide scale in recent years.

Electric powered drones are being used to deliver medical supplies in some parts of the world, for example by US-based, Zipline in Africa. Using drones for parcel delivery has been piloted and tested by Amazon and Wing in the US. Even three-minute food delivery by drone from startups like Irish company Manna is now a reality.

Piloting lightweight small-scale drones with batteries is one thing but what about electric-powered aircraft for human transport? When I was a kid and watched the Back to the Future series, I expected my DeLorean-like car to be able to fly, just powered by waste food in my portable Fusion reactor, by the time I was an adult. This was of course science fiction but drone-like passenger eVTOL aircraft are being pioneered by several startups for commercial launching.

German startup, Lilium, have developed a working five and seven-seater eVTOL short-range aircraft that can recharge just like an electric car from a standard 220/240/110-volt plug. Dubbed an all-electric air taxi in what they state will be an air mobility revolution. Echoing Tesla's philosophy, they focus on simplicity of design and user interface combined with safety. In 2019 the Lilium jet became a reality when its first maiden prototype flight was successful. Lilium's target is 300 kilometers per hour in level flight and a range of 300 kilometers. eVTOL requires high power to lift even this lightweight machine off the ground before it begins level flight. Liftoff and landing, therefore, consumes 10 times more energy than vertical level flight. To achieve this, it has rotating wings containing 36 electric ducted fan motors (EDFs). The design, inspired by a manta ray, gives a highly efficient drag ratio and flight performance. Like small electric drones, maneuverability can be controlled by individual electric fans. These fans produce low noise making this not just a sleek lightweight aircraft, however. Lillium will also be a relatively silent object in the sky. With higher battery density technology improving, it is estimated the jet will have 300 kilowatts of total stored energy, about four times a typical electric car.

It's not just new players, big established commercial airline companies, like United Airlines, are working on their eVTOL aircraft. Announcing an order for eVTOL aircraft from Californian startup Archer Aviation, United plans to fly passengers from one airport to the next over short distances for connecting flights. Planning for over 200 eVTOLs with an investment worth over $1.5 billion takes the concept from dream to serious reality. United's eVTOL is expected in 2025.

With the near exponential rate of change in the 21st century, many 20th-century business models simply won't survive in this century and no industry is immune. The big players don't plan to be left behind like the auto carmakers were by Tesla; Airbus is developing an all-electric multi-copter as an air taxi. Other startups like Joby Aviation, Japanese Skydrive and German Volocopter (a multiple rotor helicopter) aim to make hailing an eVTOL taxi as cost-effective as hiring an Uber today. Uber itself developed its UBER Elevate platform and later handed the business to Joby Aviation. This is now serious business. In Korea, over $20 billion has already been invested in air taxis, hub locations and technology. Regulatory aviation issues need to be overcome, air corridors need to be created, safety and necessary landing and takeoff hubs need to be agreed but this isn't some vague distant future, this is happening.

The American Federal Aviation Authority (FAA) has already published some concepts for future urban eVTOL operators. Once again China is trying to take the lead, with its EHang eVTOL aircraft which also focuses on delivering an autonomous experience. The global market could be worth more than a trillion dollars by 2040.

eVTOL is repeating a similar pattern to the electric car revolution. Elon Musk himself has even said he would love to do an eVTOL aircraft. From teenage dreams of flying cars to working eVTOL prototypes that are about to enter mass production, your children will almost certainly be commuting routinely in electric powered short hop aircraft.

ELECTRIC SKIES

With improvements in electric motors and battery technology, the world is now racing to build short and medium-haul planes that produce zero emissions. Promising to be cheaper to operate than fossil-fueled aircraft, electric aircraft also offer quieter skies, shorter runways and lower passenger ticket prices. Just like electric cars in the late 1800s, there were early experiments with electric airships almost 150 years ago in France. Before the age of lithium-ion, nickel-cadmium batteries simply couldn't hold enough charge given their weight to justify any form of electric aircraft other than hobby and models. Commercial aviation accounts for about 2.5% (1 billion tons per year) of global CO_2 emissions. Although this number may seem relatively small, considering cars account for ten times this number, it is still a big number, about as much as the entire South American continent emits in a year.

Worse still, air traffic is one of the most polluting forms of transport per distance travelled. 500 kilometers by aircraft is 10 to 50 times more polluting than covering the same distance in a high-speed electric train and 5 to 10 times more polluting per passenger than using a bus. For example, a car will produce 43 grams of CO_2 per kilometer travelled on average versus a domestic airline flight producing 133 grams. There is also a secondary impact from flying equivalent to another 121 grams due to high altitude flights causing increased warming effects from other non-CO_2 gases such as nitrogen oxides. Achieving sustainable aviation has been a huge focus of airlines over the last number of years as negative media attention on individuals' carbon footprints grew. Great strides have been made to improve aircraft efficiency, composite lightweight materials, more efficient engines and better fuels that have consistently lowered air pollution emissions since the 1970s. However, just like with the internal combustion engine, we are reaching the limits of what can be achieved with fossil fuel burning jets.

Solar-powered planes like the Solar Impulse 2, whose entire enormous wingspan was covered in solar panels, spent 16 months flying around the world in 2015 but with just an average speed of around 30 mph. Commercial solar planes are still farfetched with today's technology.

So, can we ever build an electric passenger plane that could also be recharged from renewable energy sources, say big solar and energy storage installations at airports for example? The answer today is partially yes. Cost and turnaround times for modern airlines is paramount as planes make no money sitting on the ground. I once watched a Ryanair plane land at Belgium's Charleroi airport, complete a full turnaround and take off again in just 40 minutes. Charging an electric plane at this speed isn't possible today. The number one issue to overcome is energy output per weight of a battery versus fuel. Right now, jet fuel contains far more punch than a battery by weight. Modern lithium-ion cells can produce about 250-Watt hours per kilogram while commercial aviation fuel is closer to 12,000 Watts per kilogram!

A 10-seater eVTOL is one thing but building a large electric commercial aircraft capable of transporting several hundred passengers over thousands of kilometers is a completely different proposition. We are quite a way off from a fully electric commercial aircraft, battery weight would need to drop at an exponential rate to deliver the same weight to power ratio and high performance as jet A1 fossil fuel. It's not just about being able to carry the weight of electric batteries, it's also about developing electric motors that are powerful enough to compete with modern jet engines. Today's jet engines are incredibly reliable and capable of enormous thrust and power, bringing heavy commercial aircraft to travel at 75% the speed of sound, up to 600 mph.

Despite these massive engineering challenges, once again the electric revolution is spilling over into the aviation world in a big way. Until battery technology evolves to enable small commercial aircraft to become a reality, there are interim solutions including hydrogen fuel cell technology. I'll cover hydrogen in the next section. Governments have been looking closely at airline CO_2 emissions. A 2021 example was when France announced it was considering banning short-haul internal airline passenger transport when an electric high-speed train route alternative exists. France produces a large proportion of its electricity from 'clean' nuclear power enabling the highly successful, fast electric TGV train.

Short-haul, medium-sized electric aircraft will certainly come before larger long-distance airliners. Long haul jumbo jet sized electric planes are quite a way off, but the good news is that roughly 45% of global passenger air traffic is less than 500 miles. Typically, these short-haul flights are serviced by smaller aircraft like a Boeing 737 or Airbus A320/319 which typically carries just 120-150 passengers. Short-haul flights produce a higher CO_2 output per kilometer than long haul because they never achieve long-distance durations at cruising altitude. High altitude cruising is more efficient than relatively short takeoff and landing cycles.

Using heavy expensive commercial aircraft, designed to travel thousands of miles, for short hops is very inefficient, however, that's what we do today. Electric short-haul flights may offer a possibility to have more and smaller airports than the giant international hubs of today. Short-haul commuter routes will be better served by smaller and more regular zero-emission electric aircraft designed for ranges of 500 miles or less. For this reason, the heavily polluting short-haul flight market is the most attractive to be the first to go fully electric. In the US, short-haul flights account for about 12% of all carbon emissions. Building a plane that could fly 500 miles or more, carry 50 to 100 passengers at speeds of 300 plus miles an hour would vastly lower operating costs - this is the prize for electric aircraft.

This reality, for smaller electric commercial aircraft, is closer than you may think. There are at least 10 startups in addition to major commercial aircraft manufacturers, like Boeing and Airbus, now working on an electric commercial plane. I'm going to give you a taste of the incredible electric experience that awaits us soon.

For over 10 years, Airbus has been working on developing a four-engine fully electric commercial aircraft, starting with the concept 'CriCri'. Using their E-Fan electric platform they have already produced a prototype aircraft, the four-rotor Vahane and E-Fan 1.1 for the inaugural 2021 electric airplane race. Airbus is also developing hybrid aircraft solutions using a combination of traditional fossil fuel jet engines and E-fans. Startups like Israeli Eviation unveiled its zero-emissions electric aircraft, 'Alice', in 2019. With almost one megawatt-hour of stored battery electricity (1,000-kilowatt-hours) Alice will be capable of a 1,000-kilometre flight (650 miles) at a cruising altitude of 10,000 feet and a speed of almost 500 kilometers an hour, (300 mph). Powered by 3 rear-facing high-performance electric propellers, Alice is targeted to be in the skies by 2024 carrying 11 people. This is just the beginning, as expected battery weight will reduce, making aircraft like the Alice and bigger variants a reality.

Commercial jet engines have had over half a century of refinement and improvement. Today these engines, manufactured by companies like GE and Rolls-Royce, are among the most efficient and reliable machines ever built. They can deliver thousands of hours of flight safely, efficiently and in the worst possible environments. High-speed electric fans and propellers have nowhere near this development pedigree. Building an electric motor capable of producing extremely high speeds in extreme conditions reliably and safely is vital if electric aircraft is to get flight certified. One startup has focused on this problem, Manix.

Indeed, the Alice is powered by three 375-horsepower magni250 motors. Like the automotive industry, Magnix can retrofit old airframes with their motors and batteries. Companies like Harbor Air Seaplanes plan to convert older aircraft with Manix Motors and batteries.

California based Ampaire has already tested its electric EEL aircraft. This plane is a converted Cessna 337 Skymaster, fitted out with a new rear electric motor and batteries while still retaining the original gas-powered motor at the front. This hybrid creation is a steppingstone to a fully electric Cessna.

In the pilot training market, companies like Bye Aerospace are developing low-cost electric training planes. Overcoming the motor and battery problem comes first, getting the world's aviation regulators to certify electric planes safe for passenger travel will be another challenge. Modern aviation is rigorously tested before certification. For these reasons, we may see old airframes retrofitted in the meantime before complete brand-new certified electric planes become a reality. With all the momentum now, building electric skies are an inevitability. The future of transportation is electric, there is no exception.

HOPELESS HYDROGEN

50 years ago, there was much hype about hydrogen being the wonder future fuel for cars, promising pollution-free motoring with just water dripping out of the tailpipe. Hydrogen is super light and highly combustible. Everyone remembers hydrogen's infamy from the doomed Hindenburg. This abundant gas is locked in our oceans, lakes and rivers, but unfortunately has no place in the future of the automobile. This doesn't mean hydrogen for transport is dead; there are use cases for heavy long-distance transportation like ships, aircraft and trucking. There have been lots of prototypes and experiments with hydrogen fuel cars and hydrogen fuel cell cars. Hydrogen fuel cars are like normal gasoline cars that burn hydrogen instead of gas while hydrogen fuel cell cars use a chemical reaction from hydrogen to produce electricity to drive a motor. In a hydrogen fuel cell, oxygen is added back into the mix to create water which in turn creates electricity

from the reaction. Despite decades of research, with billions spent and even hydrogen fuel stations popping up across the globe, it is not the right choice for passenger cars. It's too late for hydrogen, lithium-ion battery advancements have overtaken the necessity to develop hydrogen cars.

Producing hydrogen from water is a costly endeavour, requiring a huge amount of electricity. Storing and transporting hydrogen is also costly and difficult. For hydrogen to become a reality for mass transport we would effectively have to rebuild the hundreds of thousands of gas stations to become hydrogen stations. What made hydrogen attractive was that it wouldn't require any major lifestyle changes to fuel your car, you do it just the way you do with gasoline, but this is an entirely wrong comparison. The entire oil industry delivering gasoline to your pump has evolved over 100 years, it's not a simple case of replacing that liquid with hydrogen. It will be a massive infrastructure change, whereas electricity is already established across the globe.

Hydrogen is a flawed solution to climate change when it comes to cars. Hydrogen is created by electrolyzing water which cuts the H^2 out of the O in H^2O. It takes a lot of electrical energy to split the H^2 out and it's been estimated at least 30% of the energy used is lost in the process, according to studies by Transport and Environment. Once you have extracted the H^2 it then has to be stored, transported, and brought to your petrol pump which creates its carbon footprint. While you could argue that electricity also must be created to charge battery-electric cars, at least once it's generated it can travel directly via power lines to your car, unlike hydrogen where we are converting electricity inefficiently into gas then transporting that gas. There is also a further loss of hydrogen in the transportation process. It has been estimated that less than 50% of the energy from the original H^2 extraction process ends up in a car.

This compares with just a 5% energy loss from transporting electricity from a power station through power lines. Even the best hydrogen fuel cell technology loses almost half of the energy again in the process of combining oxygen with H^2 to create electricity, meaning you're only getting about a quarter of the energy output from the original hydrogen. Research and development on hydrogen fuel cells for transportation is on the wane. Only around 8,000 hydrogen fuel cell cars have been sold globally and sales of Toyota's hydrogen fuel cell car, for example, have been declining while development on hydrogen fuel cell trucks with Volkswagen's Scania has ceased.

Tesla's semi-truck, if proven successful on long haul heavy freight traffic, will likely put another nail in the coffin of hydrogen fuel cell transportation. An explosion at a chemical plant in Santa Clara, California and Norway deprived hydrogen fuel vehicles owners of their only supply. Hydrogen will likely go down in history as a short-lived and temporary distraction to the battery-powered electric revolution. During President Bush Junior's presidency, hydrogen got serious support as a future fuel for transport, but a fast track to today, the Biden administration's focus is on battery-powered transportation, with hydrogen all but forgotten.

It's unfortunate to see hydrogen's potential demise. If the technology had been proven to be more efficient and the infrastructure cheaper and quicker to deploy it could have overtaken battery-electric technology, but it's too late, the momentum is simply unstoppable towards a battery-electric future. Hydrogen fuel cell technology might linger on and survive in some niches where quick refueling is necessary. Examples include a roll-on/roll-off ferry and heavy goods trucking, as troubled manufacturer Nikola is working on, however, battery power is also making its way into shipping as I will explain in the next section. It would appear as of 2022 that the once future fuel has no future at all and without the distraction of hydrogen, the focus on a battery-electric revolution will accelerate unhampered.

ELECTRIC SEAS

From electric skies to seas, the revolution has spilt over into shipping. Global freight shipping accounts for a similar amount of CO_2 emissions, 2.5%, as the entire emissions from Germany in a year. International shipping has been steadily growing and shipping emissions could increase by 50% or even 250% without action according to an IMO Greenhouse Gases (GHG) study. The fuel used in heavy shipping is known as bunker fuel - heavy grade diesel oil. As with jet aircraft, the development of more efficient diesel shipping engines, better hull designs and even intelligent weather routing have helped to reduce CO_2 emissions. Unfortunately, these have been largely offset by ever-increasing ship sizes and traffic. Giant container ships now roam the oceans, like the goliath MV Evergreen that blocked the Suez Canal in early 2021. Freight cargo is the backbone of our modern economy, carrying 90% of world trade, and globalization means goods are shipping across the world 24/7.

Tackling the emissions from shipping is therefore an essential component of the green revolution. Even tracking the total emissions by country from shipping is fraught with difficulty, as ships are often registered in tax havens and offshore shell companies. When ships are in international waters their carbon emissions are technically regarded as international rather than originating from any one single country. This makes it difficult for regulators to control and limit pollution standards. Global shipping is not covered by the Paris Agreement because it cannot accurately be assigned to countries. If global shipping was a country, it would be the sixth biggest CO_2 emitter in the world. The answer is to tackle the ships regardless of which country is responsible. The impact of the 2020 coronavirus pandemic had just a 1% decrease in global shipping CO_2 emissions. The UN's International Maritime Organization has set a carbon reduction of at least 50% by 2050.

Given their enormous size, is it feasible to replace giant diesel engines with batteries and electric propellers and are there alternatives? What propelled ships across the globe for centuries is making a modest come back under a modern guise. High tech wind sails could be helping to propel giant cargo ships across the oceans. Companies like Wallenius Marine is developing the Oceanbird which is a partly sail-driven ship capable of carrying 7,000 cars. Even Airbus plans to equip one of its cargo ships with 'Airseas', giant sails helping to pull a ship forward. Sail assisted freight transport was experimented on during the 1970s and 80s, following the oil crisis. While sails can help, converting ships into fully electric and ultimately autonomous is underway. Already hybrid shipping with a combined diesel motor and electric batteries are in use. In Scandinavia, electric ships, which are completely battery powered are already ferrying passengers, with a 100% electric ferry now crossing between Helsingborg in Sweden and Helsingør in Denmark over 40 times a day. With short crossings, the ship charges for just 10 minutes at each end using ultra high-power chargers. The converted ferries are delivering at least a 25% efficiency gain over their former diesel incarnation with 4.1-megawatt-hours of battery storage (enough power to drive almost 25,000 kilometers in an average EV).

Just like electric cars, less maintenance is required with electric ships. Going further, cargo tankers and cruise ships are also being electrified. In China, the Yangtze River Three Gorges 1 is the largest electric cruise ship in the world with a 7.5 MW hour battery capacity. While nowhere on the scale of modern super cruise diesel ships, that can carry thousands of passengers, the Chinese effort is a start and will have the capacity for hundreds of passengers. Kongsberg & Yara are jointly developing the world's first zero-emissions fully autonomous container ship in Norway. The Birkeland will start life as a manned vessel and then transition to be fully autonomous with an 8 MW hour battery capable of 24/7 operation. In Tokyo, the 'E5' will be a battery-powered tanker and almost 60 meters long while in China ironically an electrified bulk carrier will be transporting polluting coal, for the time being. In Japan projects like the 'Wind Hunter' will be powered mostly by massive sails while beneath the hull turbines will generate electricity that is then used to run an electrolysis process to create hydrogen onboard which can then be used to power a fuel cell generator. These are just a sample of some of the many exciting electric developments in global shipping. However, the transition to a fully green electric world fleet is only in its infancy.

Even with all the benefits of reduced fuel costs, zero emissions and lower maintenance costs, batteries are still too expensive to completely replace diesel-driven giant ships. As battery costs continue to fall and energy density improves, combined with innovative new shipping technology like sails and solar cells across the ship's large surface area, there will be a gradual rather than exponential shift to electric shipping.

The era of the hybrid ship will likely last for several decades before the likes of solid-state batteries are sufficiently cheap enough to build pure electric ships. Given some of the long distances that shipping takes, it will be a while, as long-haul commercial aircraft, before these ocean traversing carriers are replaced with electric. On top of this the question of recharging at the world ports, with tight turnaround times, and the huge power loads required is still a way off.

MAAS – OWNERSHIP CAR-POCALYPSE

When I was a teenager, I was obsessed with buying my first car, my friends were equally like-minded. My only transport options growing up in Dublin were to use a bus, so the solution was to get a car, get freedom. This has utterly changed in the 21st century. Most young people don't care about owning a car especially with all the costs then go with it. Today, in most modern cities, a plethora of Mobility-as-a-Service (MAAS) alternatives exist for the non-car owner. From micro-mobility, with electric kick scooters, electric bicycles, electric mopeds, carsharing, car rental and short-term car subscription services, there are so many new ways to get around. Combining these with the phenomenon of ride-hailing services like Uber and Lyft, city dwellers today have incredible options other than just buying their car. This has directly impacted car ownership rates across the world and is expected to grow. Just like 'peak oil' what has become known as 'peak-car', referring to the percentage of individual ownership, may have been reached in Europe. A 'carpocalypse' may be underway.

Even before the Covid induced drop in car sales, personal car ownership had declined by about 10% in the United States and about 6% in Europe in 2019. A survey by the American Automobile Association found that city dwellers, particularly the Millennials and Generation Z (those aged less than 40 as of 2021) are less inclined to consider purchasing a car than previous generations. This doesn't mean global fleet numbers will drop because people still need to get around, it means that the pattern of ownership away from personal to shared fleet ownership will increase. Fleet operators seek the most cost-effective operations possible which lends nicely to an acceleration of the electric revolution. Fleet buyers will have no hope to convince financiers and bankers to purchase diesel and petrol cars from around 2024 onwards. With plummeting resale values of ICE cars, versus stable prices and a lower total cost of ownership by 2025, MaaS fleet operators will choose electric only.

Uber declared in 2020 it plans to go all-in on electric. Hertz has put in an order for 100,000 Teslas and a partnership with UK electric van and bus maker Arrival to build a specific electric car designed for its drivers. This means an entirely new generation of car users will possibly be using an electric car first, as they avail themselves of these MaaS services. For this electric generation, driving electric will be the norm and they would never consider buying an ICE car, adding more impetus to the revolution.

Another factor is that data is king. MaaS providers that operate fully electric connected fleets, capturing data and driver habits including electric car charging patterns is something that is happening today and helping to continuously refine the electric MaaS experience. In our business, UFODRIVE, we have collected hundreds of thousands of data points, anonymously, to determine the optimal electric charging experience for our customers using AI, which has ensured the elimination of range anxiety.

The future of mobility is electric, shared, connected and ultimately autonomous. There will still be a place for car ownership, the car gave us freedom and many people still like to own their car. Perhaps, as city streets become pedestrianized and the availability of MaaS services increases, car ownership may become confined to the suburbs where young families need their cars.

Ultimately many of these services may even become fully autonomous, which begs the following question: does the next generation even need to learn to drive especially as autonomous technology takes off? My daughter got her first car just a few weeks before I wrote this section of the book, she bought a manual stick shift because she wanted to learn how to use them in case she would ever need to in the future. When I was learning to drive 30 years previously, I would've laughed as a teenager at the idea of only learning to drive an automatic car, because that wasn't regarded as real driving!

Considering all cars will be electric, all electrics are automatic, and this revolution is happening ultra-fast what's the point of learning to drive a stick? Wouldn't this be like learning to drive the earliest Model T with cranks and multiple levers that we don't have today? As previous generations would marvel at the complex operation of a steam train, our grandkids will probably marvel at clutches, gears and even brakes!

The demise of car ownership will happen at different speeds across the globe and even in different places within cities. In European cities, where city dwelling is more congested than American cities, car ownership is dying at a quicker rate than in the US. Urban and suburban areas are more likely to hang onto cars a bit longer as typically MaaS services don't extend beyond cities, however, that's likely to change.

By 2025 most MaaS operators will be replacing or have already replaced their internal combustion fleet with electric. The business of a traditional car dealer, having only recently moved to online direct car sales and even used car delivery sales, has changed partly to subscription and short-term leasing models. This trend is predicted to increase with the next step being for fleet bulk buyers to abandon ICE cars completely. Shared and ultimately autonomous mobility will dominate our electric future.

AUTONOMOUS

The news headline quoted a Texas police officer stating that "There was no one in the driver's seat" following a tragic fatal Tesla car crash in which two people died in May 2021. Immediately the media jumped on Tesla's autopilot feature as the likely cause of the crash and quickly the story spread, resulting in a shock drop to Tesla's share price the next day. This is not the first time there has been a fatal crash where Tesla's self-driving technology has been denied the convention of innocence before guilt. Media scrutiny of self-driving cars, particularly Tesla, has been a feature since Elon Musk began hyping up the potential for fully self-driving vehicles. Public skepticism on self-driving cars hasn't been helped by several idiot drivers posing on social media sites sitting in the passenger seat or even rear of their vehicles while autopilot was engaged.

Full self-driving autonomous capability is not a reality yet, although it is extremely close. The current autopilot features of Tesla will control and partially navigate a vehicle on highways and city streets, but the driver must have their hands on the steering wheel. The steering wheel will detect if no hand is placed and begin an audible warning first to inform the driver to put their hands on the wheel. After this, a more forceful audible warning sounds and ultimately the vehicle will stop itself if the driver fails to place hands on the wheel. I have used autopilot many times, mostly on highways, where I leave my hand resting on the steering wheel and the car does the rest. To be completely honest I find using autopilot makes me even more aware and even stressed about the traffic around me and how the car is driving than if I was just driving it myself. This is the nub of autonomous vehicles; we simply don't trust them yet. Don't get me wrong, human error is responsible for 99% of car crashes and machines in the long term will likely make better decisions, but it's not there yet.

Following an investigation by Tesla, it turned out that the autopilot features on the fateful Texas car had not been engaged. The US auto safety agency, the NTSB, has opened almost 30 investigations into crashes of Tesla vehicles where there have been suggestions autopilot had been engaged. Despite these setbacks, including some tragic fatal accidents, autonomous vehicles have come enormously far and are almost a reality.

It's not just Tesla, there are a whole host of companies developing self-driving cars and trucks. Already small-scale autonomous robot delivery vehicles from companies like Starship Technologies, are in operation delivering parcels in urban areas. Delivery companies like Nuro are already driving along city streets autonomously. What was once the stuff of science fiction is now very much a reality and is close to becoming mainstream. Before the public will accept a machine that drives itself (even though 80% of the time when you're in a commercial passenger airliner it is flying itself on autopilot) there is a lot of proving and testing to go. I'm going to give you a high-level view of where self-driving technology is today and where it's likely to go very soon.

Autonomous vehicles, from driver assistance to fully self-driving, is defined within 5 levels. Starting with Level 0, meaning manual control with no driver assistance moving to Level 5 which is full automation. For example, Level 3 is conditional automation where the car is scanning and detecting environmental traffic situations and can perform most driving tasks, however, still requires humans to retain overwatch. This is effectively where Tesla's autopilot and most other manufacturers' autonomous technology is today. A Level 4 vehicle performs all the driving tasks, but a human override is still an option. With Level 5 there is zero human interaction required; this is the science fiction version of fully autonomous driving and where we are ultimately headed.

There are two main ways autonomous driving technology works. The first way, and Tesla's method, is to use a series of cameras on the vehicle to scan the environment to monitor everything from traffic, pedestrians, road works and even the weather. Algorithms and artificial intelligence programming have been capturing millions of miles of human driving, road conditions and traffic behaviours with Tesla's inbuilt cameras. In the case of Tesla, this is a key asset for the company and one which puts it quite far ahead in the self-driving race. The theory is that with these use cases being analyzed by the computer it will ultimately be able to make split-second decisions based on real-time environment and its database of historical driving/traffic behaviours. Other companies like Google's Waymo have been capturing similar data with its test vehicles for years.

The other method is to use radar to produce a three-dimensional view of road and traffic conditions using technologies like LIDAR (light detection and ranging). LIDAR uses a form of pulsed laser to measure exact ranges from objects as they move like cars and traffic. Some manufacturers have adopted the LIDAR method while others the camera method or a combination of both.

Going back to Tesla, its cars use machine learning so that its autopilot can distinguish between thousands of different hazards like road signs, trash cans, cyclists and the gradient of the road to make instantaneous decisions. Tesla has eight onboard cameras in total monitoring the environment constantly, even when autopilot is not engaged. These cameras cover long-range, nearfield and wide-angle views to detect objects and movement from up to 150 meters away. These cameras are designed to be highly redundant and are combined with a front-facing radar to monitor objects to confirm the visual input. Tesla also uses 12 ultrasound sensors for short-range analysis, which are wired into the autopilot system. Tesla is not just capturing data from a test fleet of self-driving cars but from every Tesla ever built and driven by every driver all over the world. Known as 'shadow mode' this monitoring has upset some over privacy issues.

Tesla has a huge self-driving data lead, and all this data is fed into its full self-driving chip, developed by Samsung, containing over 6 billion transistors. Tesla is relatively unique in its camera focused technology, while others like Ford, GM and Waymo are relying predominantly on LIDAR. The technical competition between these manufacturers has spilt over into public arguments about which technology is the best, with Elon Musk even claiming in 2020, that "LIDAR is a fool's errand" and that "anyone relying on LIDAR is doomed". Allowing drivers to completely rely on self-driving technology still requires legal approval in every country in the world and will take convincing. While Tesla came early to the autonomous party others are quickly catching up with some exciting technology.

In 2021 Volkswagen started testing its autonomous technology as part of its plans to launch commercial delivery and micro-transit services in Germany by 2025. Using software developed by US-based Argo ID, VW's plans a new generation of the iconic 1960s VW Camper style van to be known as the VW ID. Buzz electric van which will feature ID. Pilot self-driving mode.

GM plans to expand its 'SuperCruise' system to over twenty of its vehicles by 2023, covering over 200,000 miles of compatible highways and claims its technology will provide 95% hands-free driving. Ford plans to launch its self-driving capabilities in 2022. BMW and Daimler are also advancing auto-driving technologies. Uber plans to operate a fleet of 'robotaxis' soon while work on electric autonomous city busses is underway by companies such as Volvo.

Lowering transport costs with driverless trucks, buses and last-mile delivery fleets is one benefit of autonomous; safer roads is another, however, there is a climate positive potential too. Unlike humans, autonomous vehicles could be programmed to take the most efficient routes, minimize the use of acceleration, and operate 24/7. These can have energy-saving and therefore a CO_2 impact. When I use the autopilot feature on long haul highway drives in a Tesla, I do notice the battery consumption is far steadier, and lower than if I was driving myself. This is because humans will accelerate and overtake perhaps unnecessarily while the robot machine won't. Autonomous cars may be initially restricted to dedicated highway lanes or specific city routes. The possibilities of autonomous vehicles are boundless and exciting, and ultimately what we once considered science fiction is just a few years away. It will change future generations' attitude to transport and driving. Autonomous technology advancements have occurred in parallel to the rise of the electric car, driven in part by Tesla's push on autopilot and full self-driving technology but also in the advancements in artificial intelligence. High-performance cheap processors and technologies like 5G and high-speed Internet make autonomous possible. The future is electric, the future is connected, the future is largely shared and very likely fully autonomous. The path to clean electric mobility is clear, the momentum is building and now unstoppable.

So how else can the world achieve its cleaner air and carbon emissions targets? In the final chapter of the book, I'm going to cover the exciting rapid technological advancements in energy production and storage that will further fuel the electric revolution and cement a clean, bright future.

CHAPTER 10

THE GREAT GREEN TRANSITION

FIXING OUR AIR

I started this book with the story of our air and what you've been breathing from the thousands of polluting cars on your streets. I hope by now you get the message, the world is shifting electric and fast. So, if all cars, trucks, planes and boats are going electric why do we need to do anything else? Fully electric transport will greatly help reduce global CO_2 emissions, not to mention give most of us cleaner air, but it doesn't solve the climate crisis.

The world's population is expected to reach over 10 billion by 2100 – and they will not be the same as today's electricity consumers. With increasing global living standards, individual demand for electricity is going to increase tenfold. The number of cars on the road is predicted to reach two billion by 2040. Most of these cars, if not all, will be electric and have zero emissions, but the same cannot be said about their source of power unless radical international change happens. To meet this insatiable appetite for energy, we must change, and change now.

The world is going through a fundamental transition phase from the era of fossil fuels to the era of clean renewable energy and electric cars. How will we make this transition? The good news is there is hope. The great green transition is already happening and accelerating with new technologies and concepts. So far just around 10% of the world's energy is produced from renewable sources like wind, solar, hydro, wave and geothermal. Fossil fuel accounts for around 80% of our energy and nuclear just 2%. The single biggest issue with replacing fossil fuel with renewables is money.

As you will see in this final chapter many technologies already exist, are proven and can replace fossil fuel-based electricity. The problem is they are simply too expensive; the cost needs to drop dramatically before they can compete with fossil fuel. On top of this, renewables are inconsistent in their output, and therefore innovative and cheap storage solutions need to be advanced and scaled up.

Fossil fuels are simply too cheap. Gasoline is cheaper than orange juice in many countries. Global oil production from well to gas pump is well established and efficient despite the fact it's destroying our atmosphere; this can't just be replaced overnight. Irrespective of solving today's problem, developing countries deserve the same opportunity that developed countries had to enjoy a wealthier, healthier lifestyle. However, right now, they're replicating the path chosen by western countries in the last century, namely advancing with the use of fossil fuels. New coal plants are still being built around the world. This must stop immediately, but what's the alternative?

China for example, which has lifted itself from a largely rural poverty nation to an industrialized superpower, has more than tripled emissions since 1990 and even increased its emissions by almost 2% in 2020. This pattern has, and will, be repeated for almost every other developing nation unless action takes place urgently. Global fossil fuel use is accelerating, accelerating climate change. While 200 countries ratified the Paris Agreement and set targets to cut global greenhouse gas emissions, at COP 26, many are now acknowledging the potential of hitting those targets is virtually impossible. Earth's carbon dioxide atmospheric levels are at their highest level for over 3.5 million years, according to the National Oceanic and Atmospheric Administration (NOAA). Human activity and the increasing thirst for more energy, combined with population increases, means we're on a path to likely climate disaster despite the electric car revolution or advances in clean energy. The world needs to be at net zero emissions by 2050 but right now that looks like a pipe dream. This does not mean all is lost, there is still time to take corrective action. An enormous human effort is required, the incredible achievement of Covid19 vaccination development is an example of what we can do when under pressure. As the saying goes "necessity is the mother of all invention" and clean energy is a necessity. The problem is everybody needs to accept the necessity and urgency now.

To fix our air and climate, three big things need to happen. Firstly, we need to stop building fossil fuel power stations; secondly, we need to massively increase investment in new cleantech energy production to bring costs down; and thirdly, we need to create cost-effective, innovative energy storage solutions. A fourth, and less realistic move, is to advance technologies to capture carbon from the atmosphere. Just like the dropping cost of battery production has made the electric car a viable alternative to internal combustion, so too will technologies in clean energy production make them a viable alternative to fossil fuels. The key question is how quickly this can be achieved as we are running out of climate time. I'm going to explore in this final chapter of the book just how this is, can be and should be happening.

CLEAN ENERGY

Just 10% of global energy production is from renewable technology. These technologies are not entirely 'clean'. Manufacturing wind turbines and solar panels do come with a carbon cost and have great limitations. The sun doesn't always shine, and the wind doesn't always blow. Location is another issue as some countries are sunny and windier than others. You could ask why not just put millions of solar cells in the sunniest parts of the world like the southeastern United States in Arizona for example, where it averages 11 hours of sunshine per day, or in Chile in the Atacama Desert at 8.4 hours per day? The problem is this is not where the power is needed, and most electrical grids operate relatively independently. In the US for example there is no single grid given the vastness of the country, rather it is served by localized grids. This is the case even in many countries much smaller than the US.

So, putting hundreds of square miles of solar panels in Arizona isn't going to be able to power the entire US unless the cost of electricity transmission dropped twentyfold. The same is the case for wind and wave, many countries are landlocked with no access to wave power, while central continental countries have far less wind than countries with big coastlines. On top of location and power transmission, the other problem with renewables is cost. Burning fossil fuels is extremely cheap per kilowatt produced. Even if a country could produce most of its energy from wind or solar it cannot be used all at once and the sun doesn't shine at night, so storage is the answer. Today, storage per kilowatt is still far too expensive to make it a viable option on a scale basis.

This does not mean solar and wind or geothermal and hydro are never going to get there, there are massive advancements taking place, however, they cannot be treated as the solution in isolation. Nuclear power was once the great hope to power the world cleanly. I'll cover later where the world got to with nuclear and why it shouldn't be completely cast aside.

Before the turn of the century the auto industry had existed in relative technology stagnation, and so too had the energy industry. This has all changed due to the renewed focus on electric cars and of course our impending climate disaster. The smog of the 1970s that choked American and European cities has been radically improved by the introduction of regulations like America's Clean Air Act, signed by Richard Nixon in 1970, banning the burning of fossil fuel in homes and businesses in cities.

So, let's look at where we've got to so far with renewable energy, where we're going and how much it will help.

Starting with solar: almost all life on earth is powered by the sun. Solar cell technology has been around for over 50 years, but it is only in the last 10 years that photovoltaic cells have advanced in capacity and cost to make them a viable energy production solution. There is no magic behind how solar cells work, converting sunlight into power, it's a relatively simple process. Solar panels are made up mostly of silicon; when sunlight hits the silicon, electrons begin to move to create a flow of direct current electricity. This happens between two semiconductor layers known as n-type and p-type. The P layer has electrons 'missing' or 'holes'. Sunlight causes electrons to move from the n layer to the p layer, creating a current. Most of your household appliances can't use direct current, an inverter converts the DC into alternating. During the last decade prices for solar panels have reached all-time lows. The pace of advancement has the potential to make it a very economical power generation source in many countries. Globally installed solar power capacity has grown from the year 2000, with just 1,250 megawatts, to over 500,000 megawatts in 2020 or 2% of all the world's electricity (enough to power New York City for almost two months). This is expected to grow by over 50% in five years. Solar cells have become more efficient, but still today only around 25% of sunlight is converted into power in even the best cells. However, just like lithium-ion batteries, solar cells are now 10 times cheaper than in 2010.

The amount of energy the sun sends us in an hour is enough to power the world for over a year. Solar could power the entire Earth, however, right now it is not economical enough and storage is not cheap enough to make it viable. Also in densely populated small countries, like the Netherlands for example, the amount of land area required to create enough energy for the population simply wouldn't be enough. Advancements in see-through transparent solar cells have taken place, which could ultimately lead to large buildings having an electricity-generating glass for windows. There is no shortage of ideas for integrating solar, such as solar roads where solar panels are built along the medians, even floating solar cells in lakes, for cooling. These technologies are known as building-integrated photovoltaics (BIPV).

These new inventions and concepts are likely to significantly increase the percentage of solar energy over the next decade.

One of the big issues with solar panels is ironically dealing with excess heat, solar panels convert sunlight to electricity most efficiently under 25 degrees Celsius (77 degrees Fahrenheit). New approaches like Agrivoltaics, install solar panels above crop fields, to protect crops from too much direct sunlight and help cool the panels with evaporated water from the crops. Solar panels that follow the sun can also help maximize efficiency; many solar farms tilt and pivot towards the sun's rays known as photovoltaic trackers. Instead of rather ugly solar panels on rooftops, Tesla and others have developed roof tile style integrated cells to camouflage the panel into the roof.

Advancing cell technology away from silicon to newer materials is possible, for example by absorbing infrared light, letting visible light pass through, using polymers and advanced materials in the case of a window glass solar cell. Known as organic solar cells they let through about half of the light.

Once again cost is the issue, as these technologies are relatively new and still at a very high price point per kilowatt of electricity from fossil fuel. What makes them interesting is potentially adapting them for a whole set of different use cases, like the glass in your phone. Many wristwatch models have already had this technology for several years. Although these transparent solar cells only convert around 15% of the power from the sun compared to silicon's 25%, they could be installed virtually everywhere for microgeneration.

Researchers are also working on solar fabrics that can use materials integrated with solar generating cell technology, in theory, your jacket could generate power. Flexible solar panels are now also a reality that can be integrated into a different variety of structures. Producing solar panels on a massive scale is another challenge. Over half of the world's supply of polysilicon comes from just one region in China.

Like solar, wind power is getting cheaper. In the year 2000, 17,000 megawatts of global wind power were produced while in 2020 this was 600,000 megawatts, avoiding over one billion tons of CO_2 globally. Advancements in composites used to build giant wind turbines have accelerated in recent years. The use of offshore wind farms to maximize exposure to more consistent wind has increased, with several massive projects both in Europe and the US. In the UK the increasing use of wind energy has reached record levels. Wind produced energy is growing at a faster rate than solar, but the capital-intensive nature and high maintenance of wind turbines are negative.

New technologies known as solid-state wind energy have been developed where no blades or rotation is used. These technologies offer an alternative for harnessing wind power and work on the principle of converting electrostatic wind energy to electricity. The technology creates a flow of charged particles from the movement of the air by spraying fine positively charged water droplets into the airflow, creating particle movement and electricity. Prototypes of this technology have already been established in the Netherlands.

Other technology concepts like solid-state wind energy transformers work off the same principle but without water. Other concepts using wind to create vibration energy are also in development. While large wind turbines can't be installed in cities, rooftop solutions, such as the bladeless power pod from Utah based startup Halcium, offers the possibility of portable wind energy where all moving parts are contained within a pod-like cylinder. These designs dispense with the need for heavy mechanical parts and rotor blades and if successful will be game-changing microgrid options in cities.

During Easter 2021, due to a combination of sunny and windy weather, 80% of the UK's power came from renewables, 39% wind 41% solar. Renewables are not all about the sun and wind. In the US it is estimated that coastal wave energy could power almost 70% of the country. Wave and tidal power are more consistent and predictable than the sun or wind and have the potential to be even cheaper per kWh than wind or solar. Renewable energy is mostly accelerating in developed countries. As developing nations race to achieve the living standards of the west, new fossil fuel-fired electricity stations are still being built today. This is purely a cost issue, as renewables just can't compete due to intermittency and storage costs. Germany brought its last new coal power plant online as late as 2020, even though Germany plans to close all its 84 coal-fired power stations by 2038. In China, although President Xi Jinping pledged to make the country carbon neutral by 2060, the country still added almost 40 gigawatts of new coal-fired power capacity in 2020. Our fossil fuel addiction is hard to kick.

Scientists estimate we're going to run out of oil and gas before the end of the century. With increasing fossil fuel costs, bringing the cost of renewables down is vital to meet the Paris Agreement targets in any meaningful way. So, just how close are we to renewables beating the cost of fossil fuel electricity? The cost of solar and wind has reached the point where it is cheaper than coal power in many cases according to the International Renewable Energy Agency (IRENA). Solar and wind are also more efficient as fossil fuel generation has reached peak efficiency. In the 1970s a photovoltaic cell cost over $70 today it costs less than $0.50. The cost of making electricity is only part of the problem because a coal, gas or oil plant can be effectively switched on and off on demand when needed, unlike renewables. Renewables that are cheap and accessible to many is just half the problem; cheap storage is the other half.

STORAGE

Beyond batteries, there are other ways to store electricity generated from renewables at times when it's not needed. Storing energy "mechanically" is old technology. Many solutions exist that use off-peak electricity to pump water to higher levels, say at the top of a mountain, in large reservoirs and then release that water at peak times to turn turbines and generate power. This technology is known as pumped hydro, or the gravity battery, and currently accounts for over 90% of the world's energy storage.

Using batteries for grid-scale storage is not a viable cost option yet, so new technologies are necessary for mechanically storing energy and many innovations are now emerging. They are not chemical, or battery-based and have tremendous potential. But most of these storage concepts are only designed for relatively short-term use and not designed to store energy for weeks or months. Longer duration storage at the right price point is the Holy Grail and would allow renewables to become the dominant energy source.

If moving water from high elevation to low generates power, why not move something else? Companies like Energy Vault are doing just that. The company has developed an innovative take on 'hydro like' energy storage by using cranes and cables to move heavy concrete blocks up and down a vertical tower, like elevators. Power is used at off-peak times to lift heavy blocks that are then lowered at peak times to generate electricity. The company claims its solution has far less environmental impact than building giant dams and hydro pumping infrastructure.

Moving objects to create power is just one idea but tapping into thermal storage is another. The heat from renewable electricity, stored for example in large ceramic blocks, can be reused later to generate electricity at peak demands. With proper installation and storage, heat remains for days and even weeks before cooling. Normally these "heat engines" use the heat to generate steam for electricity turbines, however, new technologies that convert the heat chemically into electricity are also in the pipeline. This can be achieved using a solar cell like configuration that converts radiated heat from storage into electricity via a photovoltaic-like cell. Other companies are using excess renewable energy and converting it into compressed air tank storage. The compressed air can then be released to turn generator turbines at a peak demand time. The world's largest non-hydro storage project is in development by a Canadian company, Hydrostar, using gigawatt-scale compressed air.

Concepts like cryogenic storage are in development, where excess energy is used to cool down air to form a liquid which can then be used in a heat exchanger later to generate electricity. Other storage concepts like flywheels and even molten salts, to store heat energy, are also advancing. We have used flywheels for hundreds of years in milling, pottery, steam engines and more recently internal combustion engines. With flywheel technology excess energy is used to speed up a heavy rotating wheel or disc encased in a vacuum structure. The heavy wheel will continue to rotate for days before it needs to be 'recharged'. When peak demand electricity is needed a generator taps into the spinning wheel. Using this kind of mechanical storage dispenses with the need for big batteries and allows fast charging solutions to be installed even away from major sources of power lines. Back in the 1950s, flywheels were used for short-distance bus transportation in Switzerland. Instead of the bus containing batteries it contained a large 1,500-kilogram flywheel that spun at up to 3,000rpm to store and use energy. Called the Gyro bus, it received electricity top-ups at bus stops to keep the flywheel spinning. The flywheel was enclosed in a vacuum structure allowing it to remain spinning for up to 10 hours after its last 'charge'. Flywheel-based kinetic energy recovery systems (KERS) from companies like the Israeli Chakratec, intend to supply peak power for fast charging EV stations. A new take on an old concept.

An alternative to lithium-ion batteries is the 'Flow battery'. A flow battery stores liquid electrolyte outside of the battery structure in a positive and negative vessel. Flow batteries can be used as storage devices in 'pods' to provide hours of peak demand energy. Companies like US-based, ESS and Primus Power have already developed and supply flow battery solutions. ESS supplies its iron, salt and water battery storage product inside a portable shipping container which can be easily installed at a factory or commercial premises.

Instead of converting sunlight to electricity, converting heat from sunlight with thousands of mirrors directing the sun at a single point on a tower is in operation in several countries. The world's largest 24/7 concentrated solar thermal power plant is now in operation near Tonopah, Nevada. The 110-megawatt Crescent Dunes Solar Energy Plant, developed by US Solar Reserve, uses molten salt to store energy. Salts are pumped to the top of the tower where the sun's heat is concentrated by the thousands of mirrors and melted. These melted salts can then be stored and used immediately for energy production or later when the sun is no longer shining at night.

To date however none of these storage solutions matches the energy density of a lithium-ion battery, but that's not to say they haven't got a place in our future. Unfortunately, another issue with renewables isn't just the days when the sun or wind doesn't shine or blow but the seasons. 'Seasonal Intermittency', presents a big problem and to cope with wind and sunshine 'downtime' massive-scale energy storage solutions will be necessary. Storing this kind of seasonal energy in giant battery storage solutions isn't practical or cost-effective meaning we may see an increase in the use and development of mechanical storage in the years ahead.

NUCLEAR FUSION

Nuclear is a clean and constantly available source of energy. The problem with nuclear is that we gave up developing the technology following disastrous accidents and a loss of public confidence. What was once a scientific, wondrous form of clean energy, is now associated with danger and disaster. Three Mile Island, Chernobyl and more recently Fukushima are etched in our consciousness. Nuclear energy is perceived as dangerous, producing radioactive waste and even radioactive landscapes because of meltdowns. As a source of energy, it is clouded with misconceptions and fear. I live mostly in Luxembourg and just 20kms across the border sits the giant Cattenom Nuclear Power plant site in France. The plant's cooling towers are visible almost 30kms away. When in full operation, on a clear day, the human-created clouds are awesome and terrifying.

I can remember the Chernobyl disaster well, even living in Ireland, far from the site of the explosion. We were sent home from school early, radio and TV bulletins warned us to stay indoors as an expected radioactive cloud would drift over Europe. Indeed, it did, with levels of radioactivity increasing in mountain sheep as far away as the west of Ireland. Chernobyl contaminated almost every part of northern Europe, although mostly not at dangerous levels.

Even before Chernobyl, nuclear had a bad reputation; its demise began as early as the 1970s. Countries began turning away from the once wonder energy instead of turning back to new polluting fossil fuel stations. Germany began the process of closing all its nuclear power stations and others have declared that nuclear power has no future in their countries. Nuclear meltdowns and explosions are of course extremely rare, but the aftermaths are potentially catastrophic and long-lasting, the abandoned city of Pripyat stands testament to that. Spent radioactive fuel can remain at dangerous levels for hundreds of years and must be securely stored. Even with a bad reputation, there are virtually no polluting emissions from nuclear power. Giant nuclear plant cooling towers are emitting steam not smoke.

However, very public disasters all but ended human research and development in nuclear power. Work on smaller, more efficient nuclear reactors of course continued, however, great leaps and advancement of the technology effectively ceased since the 1980s. Had research and development continued to refine and improve the technology, nuclear power could have been far more widespread, efficient and safer. Today, just 10% of the world's power comes from over 400 nuclear reactors worldwide. By country, it varies wildly with France leading the world at 71% nuclear while in the US it is less than 20%.

However, the story of nuclear is not over. Over 50 power reactors are currently under construction across 16 countries as of 2022. The largest new nuclear reactor in Europe is expected to go live in 2022 in Finland. Nuclear energy, for the past 70 years, has been based on splitting uranium-235 and plutonium-239 atoms to create a tremendous amount of heat, to boil water, creating steam to turn turbines – this is nuclear fission. Splitting atoms creates a radioactive by-product, proximity could be fatal and even at great distances, it can still damage cells and DNA. However, there is an alternative to splitting the atom: fusing the atoms.

With nuclear fusion, lighter atoms are joined together to produce a larger one. Instead of using heavy uranium and plutonium, fusion reactors typically use hydrogen. The fusion process creates helium as a byproduct. Nuclear fusion creates the same kind of reaction as takes place in our sun. Fusion can produce three to four times greater power output than fission.

Nuclear fusion power has been talked about for over 50 years. Even as a kid I remember my science teacher in school talking about how in 30 to 40 years the world will be powered by clean fusion reactors. That hasn't happened yet. So why is nuclear fusion taking so long and is it going to be our energy saviour? The first problem is producing a controlled nuclear fusion reaction is tremendously difficult. To join the atoms together plasma needs to be heated to over 100 million degrees Celsius and then forced together using electromagnets. This reaction is very unstable and unpredictable. Achieving fusion could give us virtually limitless clean energy and many nations have been working hard to make it a reality. Maintaining a consistent high heat for the atoms to fuse to create helium has been extremely difficult as even the slightest fluctuation or turbulence in the temperature of the plasma can end the reaction process.

Nuclear fusion, unlike fission, is not a chain reaction so it cannot run away and cause a meltdown. Ensuring we can get net energy from the energy we put in to create the reaction, versus the energy that comes out from the fusion process is the goal. Solving the problem of fluctuation is getting close with the help of supercomputing models.

The International Thermonuclear Experimental Reactor (ITER) in France, has been under construction for ten years and getting close to viability. So nuclear fusion is getting closer but is still potentially decades away before it becomes commercially viable. This is too late to meet our urgent carbon emissions reduction targets, but in the second half of the century, fusion could offer us game-changing clean abundant energy.

ENERGY EVERYWHERE

Telegraph and telephone communications were made possible by extensive wired networks. Then came the mobile cell phone and our reliance on cable networks has diminished. Some developing countries skipped the wired communications era altogether and instead are mostly now relying on mobile technology. A similar situation to power generation, transmission lines and substations is likely. Advances in solar cell technology, micro-wind, geothermal, smart buildings, and even energy created from human beings, challenges the traditional power grid model. Instead of giant power stations transmitting power hundreds of miles away, what if power could, in part at least, be generated right where it's needed? This is microgeneration.

Microgeneration solutions could be everywhere and dispense with the need for big power stations and distribution lines. The most obvious example is solar panels combined with a home battery storage pack, providing self-sufficient power and even a net contribution to the grid. Integrated solar everywhere from roofs, glass and even cars opens the potential to produce a significant percentage of our future energy from microgeneration. It's not just about solar, small wind turbines and even geothermal generation could make many buildings independently powered. Tapping into the sun's energy is not just about solar cells, capturing the heat generated from the sun, known as solar thermal fuels, can capture heat in a chemical reaction to be released later for energy. Even just walking can generate power, with companies like Pavegen that have developed a floor which slightly vibrates when people walk on it generating electricity.

Combining microgeneration solutions and renewable technologies with battery storage and vehicle or device to the grid could create what is known as the 'intelligent grid'. Energy can be shared, generated, and stored from thousands of sources rather than just power stations. Internet of things (IoT) devices all connected intelligently could help load balance demand, even your smartphone could potentially feedback its power when you're not using it. On a massive city-sized scale, with millions of people, these solutions would be material. Some of these concepts are still a long way off but one is already in use with the potential to go mainstream, vehicle to grid or V2G. Not every country currently allows domestically produced power to be pumped back into the grid, and not every electric car is capable of two-way power yet. I mentioned earlier in the book Sono Motors, whose solar-powered car could in theory pump its excess energy back into your home for use. Effectively instead of having a power wall for peak power installed in your home, your car could do that job, therefore offsetting the reliance on fossil fuel power stations.

So just how mainstream are V2G solutions? If you already have solar panels or a wind turbine at your home feeding electricity back into your electricity supplier, then yes VTG is possible. With some slight modifications to your electrical supply at home, you can run power back into the grid, although the car must be capable of that first. As of 2022, not all-electric cars offer a vehicle to grid option. Tesla cars don't yet have this capability and occasionally have declined to entertain it as a possible future option. It could be argued that vehicle to the grid would compete with Tesla's own Powerwall business.
So which cars already have this technology available and what's coming next?

Bidirectional charging is possible using the CHAdeMO charging standard, contained in many electric cars like the Nissan Leaf. The cost of V2G home solutions is still relatively expensive, however, costs are expected to drop. Using your electric car to provide energy to an electricity supplier is like the energy equivalent of Airbnb. V2G solutions, therefore, offer a form of load balancing on a mass scale. This could create the concept of an energy 'grid of things'!

CARBON CAPTURE

Carbon is the basis for all life on earth. Living or dead, plants and animals store and emit carbon. However, with the expected growth in population, plus energy demands, the great electric transition will not be fast enough to lower our carbon emissions in time to avoid climate disasters. What if we could just take carbon from the atmosphere and store it?

Carbon capture technology has gained a great deal of attention in recent years for obvious reasons. So, are these futuristic dreams or near realities? There are three main carbon capture approaches, the first is at emissions source, the second from the atmosphere and the third with natural biosphere methods. At the source, methods have been used for decades to control factory emissions. They are prohibitively costly thus preventing mass use to date. Despite being around for years the technology hasn't evolved much until recently, but now many new concepts and prototype devices are in the works. Elon Musk even offered $100 million, towards a prize to come up with the best carbon capture technology.

Carbon capture, utilization and storage are known as (CCUS), has been operational in many large-scale industries since the early 1970s. The oil industry for example has been pumping carbon dioxide into the ground to help extract more oil. Selling CO_2 from air capture to oil companies to make oil wells more productive might seem counterintuitive. Known as Enhanced Oil Recovery (EOR), its use is one criticism by environmentalists of carbon capture business models. Ironically CCUS projects peaked around the year 2010 and then began a decline until more recently as investments in the technology resumed. These carbon capture technologies involve pumping excess carbon production underground in storage spaces and caves. Carbon Capture Storage (CCS) aims to trap CO_2 emissions in a liquid form before they enter the atmosphere. A conglomerate of some of the world's largest oil companies has built CCS plants in Norway capable of capturing 100,000 tons of carbon per year.

These high-cost solutions are currently massively lossmaking, however far less costly than dealing with the consequences of climate change. The International Energy Agency (IEA) estimates that CCS technologies could be capable of capturing 20% of the world's carbon dioxide emissions, however, that would require a huge international investment. CCS technology alone therefore cannot deal with our 50 billion tons of carbon dioxide emissions per year.

The second solution is to try and collect carbon directly out of the atmosphere. Direct air capture technology focuses on capturing the low concentration CO_2 from the air. These methods work by sucking in air with a big fan, then using chemicals to extract and concentrate the CO_2 to produce solid, often calcium-based, pellets which can be stored or mixed in products like concrete.

Big oil companies, like Exxon, have formed partnerships to advance technologies to try and bring large scale breakthroughs that can remove carbon dioxide. Captured carbon could then even be stored in empty, former oil fields underground. The cost of air capture is more than fifty times the cost of a natural biologic solution like tree planning. Drawing down carbon from the atmosphere is still a worthy endeavour with startups like SeeO2, Carbon-free, Carbon engineering, Carbonworks and Deep Branch all pioneering differing technologies.

Sequestering atmospheric carbon can be done by human invention or the natural way by using the Earth's own carbon storage technology – by growing stuff! This is known as biological carbon capture. Many companies use tree planting projects to offset their carbon emissions. Trees, grass, wetlands, the sea, and the soil absorb carbon dioxide.

There are many other carbon issues we need to tackle, for example, protecting our topsoil (which stores massive quantities of carbon). Modern farming and tilling techniques are stripping our topsoil bare and releasing millions of tons of carbon, while organic methods of farming help reverse this. Protecting forests, wetlands, mangroves, river deltas and bogs also prevent carbon release. Reducing meat consumption and developing plant-based meat substitutes in addition to reducing food waste are also important steps.

As I mentioned at the beginning of the book, the planet can self-heal and enabling biodiversity to return will help to offset our negative climate activities. Tree planting helps the problem, however, we can never plant enough trees to sequester all the carbon in the atmosphere. Worse still, we are continuing to deforest huge areas, with over 10% of the Amazon lost in the last 10 years.

All these actions, including carbon drawdown, are now a necessity as simply just lowering our carbon emissions is not enough to limit warming. Accelerating these initiatives takes money, lots of money and the good news is that the investment world and governments are waking up to the challenge and opportunity.

GREEN RUSH

Solving our climate crisis should be regarded as one of the greatest investment opportunities since the gold rush or age of oil. We are at the beginning of a great green revolution and the opportunities are boundless. Instead of worrying about jobs and skills being lost as the car industry transitions to electric or the oil industry transitions to renewables, we should be looking at the massive opportunities for new employment, new skills, and progress.

Billions of dollars are now beginning to pour into this green rush. Large investment firms, pension trusts, venture capital and private equity have launched large ESG (Environmental, Social and Governance) and green technology funds. Firms like US Blackrock, the largest investment firm in the world, announced a $600 million fund in early 2021 to invest in firms working to reduce carbon emissions. Funds dedicated to green tech investments have surged more than 50% to over $1.3 trillion. Presidents Biden's administration's doubling of carbon saving targets, to a 50% reduction by 2030, opens the potential for the green rush to become a stampede of investment. Climate Capital is now big business and going to get much bigger. It's not just investment funds, corporations are getting in on the action too. Microsoft, for example, announced in early 2021 that they are launching a $1 billion climate innovation fund to accelerate the development of carbon reduction and removal technologies.

Climate capital is a 'mega-trend' gaining momentum. More and more companies are looking for a more sustainable operating model. Growing public attention on those companies that do not meet green goals or have strong green initiatives could see customers turn away. Corporations are not blind to this and what was once lip service is now becoming real investment intent. What was normal and accepted in the past may not be accepted for much longer; investing in hydrocarbon-based companies was once sound investment advice, but their ability to produce multiples of return is waning as hotter, more exciting cleantech alternatives emerge. Firms like BlackRock have declared that oil and gas investments may not be good ideas going forward. As the saying goes 'follow the money.'

Achieving 100% electric cars and clean energy is an enormous task but not impossible. The progress in the first fifth of this century has been staggering.

Our air and climate future may be a lot cleaner and brighter.

Good news for us all.

EPILOGUE

My journey began with a simple test drive, curiosity about this new electric machine. Almost five years passed since I first test drove an electric car, a Tesla before I began to write this book. When you embark on something completely new, outside of your life experience, even in your late 40s, it's only after you look back that you realize how much you just didn't know and were oblivious to.

I drove big, heavy SUV polluting gas guzzlers for years, diesel ones. Ironically, I am big into the outdoors and in particular cycling, and I was always aware of that smell of exhaust fumes in busy streets and even country roads but never really gave it a second thought.

I never, until now, considered my SUV gas guzzler days, I simply accepted air pollution and climate change. This was an abstract problem that I knew little about or considered something that I could influence.

Buying an electric car was the first step in my journey, quitting my corporate life to create an electric mobility company was the second, and everything I learned since I began researching and writing this book was the third. This book started as a guide for all those people I have met that are as oblivious to the electric revolution and its implications as I was. All those people I met who made statements based on misinformation and lack of knowledge about electric cars and this great green transition we are living through.

Even before I began writing this book, I thought I had most of the information I needed to tell the story of the electric revolution. However, the deeper I went, and the more I learned, the more horrified I became about the utter destruction we have done to our planet yet also the more hopeful I became about the solutions we have available to undo it. Researching this book also led to a great deal of frustration that solutions to many of our problems exist today, however, are held back by vested interests, lies, a lack of government will, and money being directed to the wrong places.

What I learned along the way was just how poisonous and dangerous the air I breathe is to me and you. What I have learned is that this air kills seven million people per year, far more than our disastrous Covid pandemic. What I learned is that Covid has been made much worse by air pollution and those who have suffered the most are often in the poorest and most deprived regions of the Earth. What I've learned is that these regions have the most to lose from climate change, but as they strive, rightfully so, for better living standards that developed countries enjoy, the damage to our climate will accelerate massively and beyond the point of repair. We are at a crucial stage in human civilization's development.

There is a theory called 'The Great Filter', in which it is theorized that if civilizations have developed throughout the universe, then they likely hit 'survival tests' along the way. These tests could be disease, overpopulation, lack of resources, natural disasters, or even extinction-level events like an asteroid hit. The dinosaurs didn't make it through the Great Filter. We may have already passed and survived a few tests already, like global nuclear war, but we may not survive the planet overheating due to climate change because of carbon emissions from our 200 years of fossil fuel burning frenzy. What I have learned is that we are in this situation in part because of lies and cover-ups from the hydrocarbon industry, internal combustion automakers and vested fossil fuel interests. We are of course all to blame, consuming at will and without thought for consequence. However, there is good news and a lot of hope.

The electric car had a brief life at the turn of the last century only to be replaced by the wrong technology. Our obsession with the car, with over a billion internal combustion engines in operation today, has given the world great freedom and mobility, but this era has been a period of technological stagnation. Since the adoption of internal combustion, the level of advancement has been pedestrian with just a threefold improvement in efficiency and performance since the Ford Model T.

Over these past five years of writing this book, I have zero doubt that we are living in the middle of a rapidly accelerating electric revolution. The momentum is now unstoppable and we're past the tipping point. The age of internal combustion and ultimately the age of oil is at the beginning of its end. What started as a novel concept became the most valuable car company in the world, Tesla, and Elon Musk, regardless of your opinion of him, has turned the world of the automotive industry upside down. If there ever needed to be a dictionary definition of the word "disruption" Tesla could be it. It is almost 20 years since the founding of Tesla, and for the first 10 years it was just a curiosity and partial annoyance to the legacy auto industry, however, that has changed in recent years. Even over the months since I've been writing this book, the acceleration towards an electric generation has increased more and more. Almost every major automaker has now declared its intention to become electric.

If you have any doubts or believe that electric is not for you then think again, this would be equivalent to saying you would never buy a mobile phone 20 years ago. You will almost certainly be driving an electric car within the next five years and then forevermore. The story of the rise of Tesla and its fundamental change to the way we think about electric cars and mobility will go down in history as one of the driving forces behind our third great industrial revolution. In speaking with hundreds of people over these years I realized the basic concepts around electric cars, facts or fiction are completely mixed up and most people had simply no idea about the viability, experience, and future of electric mobility.

What I have learned is that there is a general fear of change when it comes to the necessary steps to beat climate change. This fear is in part justified, as jobs will be lost and industries will disappear, but also this fear is very much due to deliberate misinformation, deliberate lies, and deceitful efforts by vested interests.

The technology of the past century is no longer viable in this century. In the first 20 years of this century, we have massively increased our population, our consumption and climate pollution, however, we've also advanced technologies that would have seemed science fiction in the 1990s. Big oil money and our complacency have compounded the problem for decades but the realization that we are now at the beginning of the end of the oil age has hit. From the Paris Agreement to individual action and a new corporate consciousness around climate change, momentum to a bright clean future is accelerating. What I have learned is that there are boundless opportunities in this great green transition, and that huge technological leaps forward have already been made and are rapidly advancing.

Ways to produce clean energy from new renewable technology, ways to store this clean energy for later use and ways to attempt to repair the carbon emission damage of the past are already well advanced. The fossil fuel industry is just waking up to its obligation and to the opportunity the green transition represents. Instead of fearing this change, criticizing new technologies like electric cars, and turning a blind eye to the damage we are causing, everyone should embrace the potential for our green future. What I've learned is that the electric car revolution has spilt over into lots of other mobility solutions, energy technologies and even financial investing.

Now that I am aware of the horror of what we've done to our planet, the blatant abuse of our resources and the world my five children, and potentially lots of grandchildren, will inherit, I know I and everyone else can influence change. I know this change is extremely positive and presents incredible opportunities for everybody. Despite all the negative things I've learned, I am filled with hope about the potential of human ingenuity and ideas combined with passion and the necessity to solve our future.

The electric revolution is a tremendous chance to rebuild manufacturing, retrain workers and begin reversing our climate damage. The future is bright. On a clean fresh, bright sunny day, sometime later this century, your grandchildren will walk down a city street and as they walk, near-silent, clean, autonomous transport vehicles will drive past them. The glass on the buildings will be converting sunlight into energy, micro static wind generation will be installed on rooftops, every store, office building and restaurant will be near energy self-sustainable and not relying on legacy power plants.

Most of the electricity being used will be coming from solar, wind or even possibly clean nuclear fusion and all of it was stored in innovative, low carbon clean technologies. The stories from their grandparents about how streets were filled with poisoned air from inefficient fossil fuel burning machines and how the lights were powered by burning coal and oil, threatening the survival of life on our planet, will seem like the darkest of the dark ages to them.

This will be the story of our second chance and how the electric vehicle rebirth was the catalyst that began our great green transition. Your grandchildren will be the green, clean electric generation. Making this happen will require huge government effort, individual actions, and a big change in corporate activity. Everything from a carbon tax to cleantech research investment to the education of the individual will be required. These actions must happen fast as we are simply running out of time.

Cleaning our air and averting climate disaster is going to take all human ingenuity to solve. The electric car is a big part of the solution, not just for the emissions cars produce themselves, but for the car's part in spurring increased attention on our sources of electricity and renewable technology. We are living in the third industrial revolution as we transition from fossil fuels and internal combustion to renewables and electric cars. We are gradually leaving the energy and mobility dark ages and embarking on a new bright future, and it can be done with relentless focus and the sheer human will to survive. We have reached another test in the Great Filter. When we truly commit to a goal, humans can achieve anything. This was never more demonstrated than by JFK in his 1962 speech:

"We choose to go to the Moon in this decade and do the other things, not because they are easy, however, because they are hard; because that goal will serve to organize and measure the best of our energies and skills, because that challenge is one that we are willing to accept, one we are unwilling to postpone, and one we intend to win"

Seven years later men walked on the moon.

We are at a pivotal moment in human history as we move from the energy of the past to the energy of the future. Our next generation is an electric one. They will have electric mobility, they will have clean air and they will witness the transition to clean energy. Where there's a will there's always a way.

ACKNOWLEDGEMENTS

Having the idea to write a book is one thing, turning it into a reality is another! It has been at times a very demanding task but one which I am grateful for the opportunity to have had.

Becoming immersed in the world of electric cars, cleantech and climate change over these past years inspired me to bring all the conversations and insights I have gained into a book. None of which would have been possible without the support of my colleagues, business partners, family and friends. The process of building the book brought me to previously unknown areas I never knew existed. It is only having completed the book that I realise now how little I knew when I started out! The new, and much-needed world, of clean mobility, technology and new energy is filled with inspiring enthusiastic people.

The learnings and contributions have come from the hundreds of individuals I have met on my startup journey. A diverse group of wonderful people from carmakers, climate scientists, entrepreneurs, investors and customers have all in part contributed to this book. Ultimately this book came about because of passion for the planet and new technology leading to the creation of startup UFODRIVE.com. There are too many to give fair acknowledgement to however I would like to give special thanks to those who put their trust in me from the beginning. In particular, Renaud Marquet, UFODRIVE co-founder, business partner and friend for over ten years. Stephen Morrissey and Stany Blanvalet for taking the plunge into the startup world when we had just an idea. Thank you to Stephen Maguire and David Yu for helping me finalise the book. To all our partners who supported us in the early days, René Steinhaus (CEO Luxembourg Airport) for our first location, Chairman John Collins for trusted guidance from the beginning and of course to our first investors for taking the bet on us!

During the writing of this book, I interviewed experts in their field and met many established executives and entrepreneurs from the world of electric cars, cleantech and new mobility. Sincere gratitude to those who devoted time to me in researching this book and providing insights and opinions. This is not an exhaustive list, however, special thanks to the many inspiring individuals I met at Tesla Motors, Hyundai Motors, GM, Cleantechnica, Volkswagen, Interparking, Uber, J.P.Morgan, Chargepoint, CNET, Eneco, Siemens, Goodyear, Sono Motors, JuiceBar, QPark, Electrive, Electrek, Transport & Environment, The European Commission for Sustainable Urban Mobility and The European Investment Bank.

Sincere gratitude to my family for inspiring, encouraging and putting up with me! To my parents Roddy and Kay, for the years of support. To my friends for the advice and encouragement all along the way.

INDEX

A

B

C

D

E

F

G

H

I

J

K

L

M

N

O

P

R

S

T

U

V

W

X

Z

Printed in Great Britain
by Amazon

86518634R00176